Anatomy & Physiology Workbook For Dummies®

Latin and Greek Roots

English Form	Meaning	Example	English Form	Meaning	Example
angi(o)–	vessel	angiogram	hemat(o)–	blood	hematology
arthr(o)–	joint	arthritis	hist(o)–	webbing (tissue)	histology
bronch–	air passage	bronchitis	hyster(o)–	womb	hysterectomy
calc(i)–	calcium	calcify	lig–	to bind	ligament
card(i)–	heart	cardiovascular	osteo–	bone	osteoblast
cili–	small hair	cilia	pleur–	side, rib	pleural cavity
corp–	body	corpus luteum	pulm(o)–	lung	pulmonary
crani–	skull	cranium	ren–	kidney	renal
cut(an)–	skin	cutaneous	squam–	scale, flat	squamous
gastr(o)–	stomach, belly	gastric	thorac–	chest	thoracic
gluc(o)–	sweet, sugar	glucosa	vasc–	vessel	vascular

Latin and Greek Prefixes and Suffixes

English Form	Meaning	Example	English Form	Meaning	Example
a(n)–	without, not	anaerobic	iso–	equal, same	isotope
aut(o)–	self	autonomic	meta–	beside, after	metacarpus
dys–	bad, disordered	dysplasia	ortho–	straight, correct	orthopedic
ec–, ex(o)–, ect–	out, outside	exoskeleton	para–	beside, near, alongside	parathyroid
end(o)–	within, inside, inner	endometrium	peri–	around	pericardium
epi–	over, above	epidermis	sub–	under	subcutaneous
hyper–	excessive, high	hyperextension	trans–	across, beyond, through	transplant
hypo–	deficient, below	hypothalamus	–blast	to sprout, to make, to bud	chloroblast
inter–	between, among	interoceptor	–clast	to break, broken	osteoclast
intra–	within, inside	intraocular	–crine	to release, to secrete	endocrine

Anatomic Cavities

- **Ventral cavity:** Extends from just under the chin to the pelvic area, encompassing the thoracic cavity, diaphragm, and abdomino-pelvic cavity
- **Thoracic cavity:** Contains the heart and lungs
- **Abdomino-pelvic cavity:** Contains the organs of the abdomen and pelvis

- **Spinal cavity:** Enfolds and protects the spinal cord
- **Cranial cavity:** Inside the skull and enclosing the human brain
- **Dorsal cavity:** Contains posterior body organs extending from the cranial cavity into the vertebral canal housing the spinal cord

For Dummies: Bestselling Book Series for Beginners

Anatomic Positions

- **Anterior:** Front, or toward the front
- **Posterior:** Back, or toward the back
- **Dorsal:** Back, or toward the back (think of a whale's dorsal fin)
- **Ventral:** Front, or toward the front (think of an air vent)
- **Lateral:** On the side, or toward the side
- **Medial/median:** Middle, or toward the middle
- **Proximal:** Nearer to the point of attachment (such as the armpit)
- **Distal:** Farther from the point of attachment
- **Superior:** Situated above, or higher than, another body part
- **Inferior:** Situated below, or lower than, another body part
- **Peripheral:** Away from the center

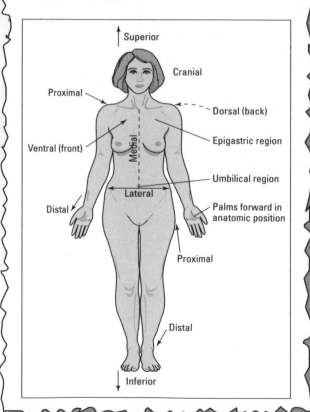

Anatomic Planes

- **Frontal or coronal:** Divides the body into front (anterior) and back (posterior)
- **Sagittal or median:** Divides the body lengthwise into right and left sections
- **Transverse or horizontal:** Divides the body horizontally into top and bottom sections

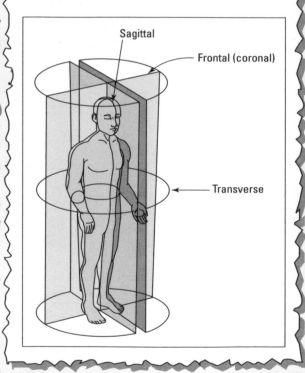

LifeART Image Copyright © 2007. Wolters Kluwer Health - Lippincott Williams & Wilkins

For Dummies: Bestselling Book Series for Beginners

Anatomy & Physiology Workbook

FOR DUMMIES®

by Janet Rae-Dupree
and Pat DuPree

BICENTENNIAL
1807
WILEY
2007
BICENTENNIAL

Wiley Publishing, Inc.

Anatomy & Physiology Workbook For Dummies®

Published by
Wiley Publishing, Inc.
111 River St.
Hoboken, NJ 07030-5774
www.wiley.com

For general information on our other products and services, please contact our Customer Care Department within the U.S. at 800-762-2974, outside the U.S. at 317-572-3993, or fax 317-572-4002.

For technical support, please visit www.wiley.com/techsupport.

Wiley also publishes its books in a variety of electronic formats. Some content that appears in print may not be available in electronic books.

Library of Congress Control Number: 2007932378

ISBN: 978-0-470-16932-2

Manufactured in the United States of America

10 9 8 7 6 5 4

WILEY

About the Authors

Janet Rae-Dupree has been covering science and technology in Silicon Valley since 1993 for a number of publications, including *U.S. News & World Report, BusinessWeek,* the *San Jose Mercury News,* and the *Silicon Valley/San Jose Business Journal.* She was a frequent guest on cable channel Tech TV's "Silicon Spin" technology talk show, and she was part of the Pulitzer Prize-winning team at the Los Angeles Times covering the city's riots in 1992. During the 2005–2006 academic year, Janet was a John S. Knight Journalism Fellow at Stanford University, where she studied human biology and researched innovation and market transfer. She freelances for various publications, working from her home in Half Moon Bay, California, where she lives with husband, Dave Dupree, and their 9-year-old son, Matthew (although a 19-year-old calico cat named Trillian actually rules the roost).

Pat DuPree taught anatomy/physiology, biology, medical terminology, and environmental science for 24 years at several colleges and universities in Los Angeles County. She holds two undergraduate life science degrees and a master's degree from Auburn University and conducted cancer research at Southern Research Institute in Birmingham, Alabama, before joining the Muscogee Health Department in Columbus, Georgia. In 1970, she moved to Redondo Beach, California, where she was a university instructor and raised her two sons, Dave Dupree and Mark DuPree. Now Pat is retired and lives on lovely Pine Lake in rural Georgia with her husband, Dr. James E. DuPree.

Dedication

To our loving family — Dave Dupree, Matthew Dupree, and Jim DuPree — for their patience and understanding as we pulled this book together and put so many other aspects of our lives on hold. To Dave, for being Mr. Fix-It on balky hard drives, recalcitrant computers, and fussy printers; for distracting an antsy 8-year-old with baseball, movies, and games; and for concocting terrific — and healthy! — home-cooked meals. To Matthew, for understanding why Mommy couldn't drive for every school field trip, attend every Cub Scout den meeting, or set up play dates every single day of the week.

And especially from Pat to Jim for his love, enthusiastic support, assistance, and encouragement without which she could not have finished this workbook.

Authors' Acknowledgments

We owe gratitude to so many people, but this project has been first and foremost a DuPree family affair. We will never be able to thank our spouses enough for their patience and understanding from beginning to end. Dr. James E. DuPree was a constant helpmate to Pat on the East coast, providing research, editing, computer assistance, and graphics support as well as priceless cheerleading and enthusiasm. Their son David E. Dupree filled the cheerleader role on the West coast, keeping the computers and home network up and running, Janet focused and on track, and the gaping child-rearing gaps filled while she pushed to get the book done in what surely must be record time. And we can't forget to thank son/grandson Matthew J. Dupree for his stalwart patience while Mom and Grandmom focused their attentions on the book instead of him.

We would also like to thank our agent, Matt Wagner, for his tireless efforts, as well as the many devoted people at Wiley Publishing, particularly Stacy Kennedy, Chrissy Guthrie (who welcomed Sophia Rose Guthrie into the world days after receiving the final manuscript!), Stephen Clark, and Elizabeth Rea.

Publisher's Acknowledgments

We're proud of this book; please send us your comments through our Dummies online registration form located at www.dummies.com/register/.

Some of the people who helped bring this book to market include the following:

Acquisitions, Editorial, and Media Development

Senior Project Editor: Christina Guthrie

Project Editor: Stephen R. Clark

Acquisitions Editor: Stacy Kennedy

Senior Copy Editor: Elizabeth Rea

Technical Editor: Michael W. Pratt, PC

Editorial Manager: Christine Meloy Beck

Editorial Assistants: Erin Calligan Mooney, Joe Niesen

Cover Photos: © Robin Lynne Gibson/Riser/Getty Images

Cartoons: Rich Tennant (www.the5thwave.com)

Composition Services

Project Coordinator: Patrick Redmond

Layout and Graphics: Stacie Brooks, Carrie A. Cesavice, Denny Hager, Stephanie D. Jumper, Julie Trippetti, Erin Zeltner

Special Art: Kathryn Born, M.A.

Anniversary Logo Design: Richard Pacifico

Proofreaders: Broccoli Information Management, Cynthia Fields, John Greenough

Indexer: Sherry Massey

Publishing and Editorial for Consumer Dummies

> **Diane Graves Steele,** Vice President and Publisher, Consumer Dummies

> **Joyce Pepple,** Acquisitions Director, Consumer Dummies

> **Kristin A. Cocks,** Product Development Director, Consumer Dummies

> **Michael Spring,** Vice President and Publisher, Travel

> **Kelly Regan,** Editorial Director, Travel

Publishing for Technology Dummies

> **Andy Cummings,** Vice President and Publisher, Dummies Technology/General User

Composition Services

> **Gerry Fahey,** Vice President of Production Services

> **Debbie Stailey,** Director of Composition Services

Contents at a Glance

Table of Contents

Introduction

*W*hether your aim is to become a physical therapist or a pharmacist, a doctor or an acupuncturist, a nutritionist or a personal trainer, a registered nurse or a paramedic, a parent or simply a healthy human being — your efforts have to be based on a good understanding of anatomy and physiology. But knowing that the knee bone connects to the thigh bone (or does it?) is just the tip of the iceberg. In *Anatomy & Physiology Workbook For Dummies,* you discover intricacies that will leave you agog with wonder. The human body is a miraculous biological machine capable of growing, interacting with the world, and even reproducing despite any number of environmental odds stacked against it. Understanding how the body's interlaced systems accomplish these feats requires a close look at everything from chemistry to structural mechanics.

Early anatomists relied on dissections to study the human body, which is why the Greek word *anatomia* means "to cut up or dissect." Anatomical references have been found in Egypt dating back to 1600 BC, but it was the Greeks — Hippocrates, in particular — who first dissected bodies for medical study around 420 BC. That's why more than two millennia later we still use words based on Greek and Latin roots to identify anatomical structures.

That's also part of the reason so much of the study of anatomy and physiology feels like learning a foreign language. Truth be told, you *are* working with a foreign language, but it's the language of you and the one body you're ever going to have.

About This Book

This workbook isn't meant to replace a textbook, and it's certainly not meant to replace going to an actual anatomy and physiology class. It works best as a supplement to your ongoing education and as a study aid in prepping for exams. That's why we give you insight into what your instructor most likely will emphasize as you move from one body system or structure to the next.

Your coursework most likely will cover things in a different order than we've chosen for this book. We encourage you to take full advantage of the table of contents and the index to find the material addressed in your class. Whatever you do, certainly don't feel obligated to go through this workbook in any particular order. However, please do answer the practice questions and check the answers at the end of each chapter because, in addition to answers, we clarify why the right answer is the right answer and why the other answers are incorrect; we also provide you with memory tools and other tips whenever possible.

Conventions Used in This Book

Half the battle of studying anatomy and physiology is getting comfortable with the jargon. To help you navigate through this book, we use the following typographical conventions:

- *Italics* are used for emphasis and to highlight new words or terms that are defined in the text.
- **Boldface** is used to indicate keywords in bulleted lists or the action parts of numbered steps.
- `Monofont` is used for Web site addresses.

Foolish Assumptions

In writing *Anatomy & Physiology Workbook For Dummies,* we had to make some assumptions about you, the reader. If any of the following apply, this book's for you:

- You're an advanced high school student or college student trying to puzzle out anatomy and physiology for the first time.

- You're a student at any level who's returning to the topic after some time away, and you need some refreshing.

- You're facing an anatomy and physiology exam and want a good study tool to ensure that you have a firm grasp of the topic.

Because this is a workbook, we had to limit our exposition of each and every topic so that we could include lots of practice questions to keep you guessing. (Believe us, we could go on forever about this anatomy and physiology stuff!) In leaving out some of the explanation of the topics covered in this book, we assume that you're not just looking to dabble in anatomy and physiology and therefore have access to at least one textbook on the subject.

How This Book Is Organized

Anatomy and physiology are very far-reaching topics, so it only makes sense that this workbook is divided into parts, each of which is divided into a number of chapters. The following sections preview the part topics to give you an idea of what you can find where.

Part 1: Building Blocks of the Body

We begin at the very beginning — chemistry — because it's a very good place to start. Stop moaning and groaning — chemistry really isn't as difficult as it's been made out to be. It's an integral part of understanding what the body's cells are doing and how they're doing it. We cover the basics from the atom on up and introduce the processes that keep the whole package operating smoothly. Then we take a look at the cells and the tiny structures inside every one. Cells are living things, just like the bodies of which they are a part, and they have the same cycles as all living things do: They grow, mature, reproduce, and die. By layering thousands upon thousands of similar cells on top of one another, tissues with unique structures and functions are formed. This part covers the primary types of tissues and where you'll find them in the body.

Part II: Weaving It Together: Bones, Muscles, and Skin

The chapters in this part have a kind of classic leg-bone-connected-to-the-hip-bone feel to them — what people generally think of when they hear "anatomy and physiology." We take you on a tour of the skeleton and then attach muscles to that skeleton to get the whole package moving and grooving. Then we wrap it all together in the body's largest single organ: the skin.

Part III: Feed and Fuel: Supply and Transport

No man is an island, and no one can exist without a consistent supply of life's little necessities. In this part, we breathe life into the respiratory system with a close look at the lungs and everything attached to them, we feed your hunger for knowledge about how nutrients fuel the anatomical package, and we get to the heart of the well-oiled human machine to show how the central pump is the hardest-working muscle in the entire body. None of that matters without a strong defense system, so we touch on the lymphatic system. And don't forget: All that metabolizing is bound to lead to some waste and by-products; we package up the trash and show you how the body takes it to the dumpster.

Part IV: Survival of the Species

No, we don't talk about who's going to get voted off the island. But we do take a close look at perpetuating humanity through reproductive successes. This part takes the male and female halves of the equation one at a time, delving into the parts of the male and female reproductive systems as well as the functions of those parts. We cover sperm and eggs, but we don't even try to address which came first!

Part V: Mission Control: All Systems Go

We have lift off! We already have gotten things moving before this part, but now it's time to study how nerves and hormones keep things hopping. In this part, we lay out the basic building blocks of the nervous system, help you wire it all together, and then show you how the body sends messages flying along a solid spine of brainy material. After that, we come to our senses with an overview of the eyes and ears (we cover taste in the digestive system chapter, touch in the skin chapter, and smell in the respiratory chapter). Then we turn hormonal to absorb what the endocrine system does, including observing the functions of the ringmaster of this multi-ring circus, the pituitary gland. We also delve into the various hormones coursing through your body, why they're there, and how they do what they do.

Part VI: The Part of Tens

This part is a classic *For Dummies* feature — lists of ten things that we just have to share with our readers. First we identify ten Web sites that can help you advance your knowledge of anatomy and physiology. Then we give you a list of ten key things to keep in mind as you study this illustrious and fascinating topic.

Icons Used in This Book

Throughout this book, you'll find symbols in the margins that highlight critical ideas and information. Here's what they mean:

The tip icon gives you juicy tidbits about how best to remember tricky terms or concepts in anatomy and physiology. It also highlights helpful strategies for fast translation and understanding.

The example icon marks questions for you to try your hand at. We give you the answer straightaway to get your juices flowing and your brain warmed up for more practice questions.

The remember icon highlights key material that you should pay extra attention to in order to keep everything straight.

The sizzling bomb icon — otherwise known as the warning icon — points out areas and topics where common pitfalls can lead you astray.

Where to Go from Here

If you purchased this book and you're already partway through an anatomy and physiology class, check the table of contents and zoom ahead to whichever segment your instructor is covering currently. When you have a few spare minutes, review the chapters that address topics your class already has covered. It's an excellent way to prep for a midterm or final exam. If you haven't yet started an anatomy and physiology class, you have the freedom to start wherever you like (although we suggest that you begin with Chapter 1) and proceed onward and upward through the glorious machine that is the human body!

Part I
Building Blocks of the Body

The 5th Wave   By Rich Tennant

"Who wants to help Grandma make her famous gingerbread man cookies? You kids get the flour, eggs, and sugar, and I'll get the protoplasm and epithelial tissue."

In this part . . .

Before beginning to study the parts of the body, it's important to know about the basic building blocks and functions that make those parts what they are. That means getting down to the true basics: chemistry, cells, cell division, and how tissues are formed. We know your eyes are glazing over, but it's really not as bad as you may think.

This part helps you discover that chemistry isn't all that tough, particularly when you focus on the organic elements involved in the chemistry of life. You look at how that chemistry takes place inside the bricks-and-mortar of the body (its cells) and take things a step further with the wonders of self-perpetuation through mitotic cell division.

Chapter 1

The Chemistry of Life

*W*e can hear your cries of alarm. You thought you were getting ready to learn about the knee bone connecting to the thigh bone. How in the heck does that involve (horrors!) *chemistry?* As much as you may not want to admit it, chemistry — particularly *organic chemistry,* or that branch of the field that focuses on carbon-based molecules — is a crucial starting point for understanding how the human body works. When all is said and done, the universe boils down to two fundamental components: *matter,* which occupies space and has mass; and *energy,* or the ability to do work or create change. This is the chapter where we review the interactions between matter and energy to give you some insight into what you need to know to ace those early-term tests.

Building from Scratch: Atoms and Elements

All matter — be it solid, liquid, or gas — is composed of atoms. An *atom* is the smallest unit of matter capable of retaining the identity of an element during a chemical reaction. An *element* is a substance that can't be broken down into simpler substances by normal chemical reactions. There are 92 naturally occurring atoms in nature and 17 (at last count) artificially created atoms for a total of 109 known atoms. However, additional spaces have yet to be filled in on the periodic chart of elements, which organizes all the elements by name, symbol, atomic weight, and atomic number. The key elements of interest to students of anatomy and physiology are

- ✔ Hydrogen, symbol H

- ✔ Oxygen, symbol O

- ✔ Nitrogen, symbol N

- ✔ Carbon, symbol C

HONC your horn for the four organic elements. These four elements make up 96 percent of all living material.

Atoms are made up of the subatomic particles *protons* and *neutrons,* which are in the atom's *nucleus,* and clouds of *electrons* orbiting the nucleus. The *atomic weight,* or *mass,* of an atom is the total number of protons and neutrons in its nucleus. The *atomic number* of an atom is its number of protons or electrons; conveniently, atoms always have the same number of protons as electrons, which means that an atom is always electrically neutral because it always has the same number of positive charges as negative charges. Opposite charges attract, so negatively charged electrons are attracted to positively charged protons. The attraction holds electrons in orbits outside the nucleus. The more protons there are in the nucleus, the stronger the atom's positive charge is and the more electrons it can attract.

Electrons circle an atom's nucleus at different energy levels, also known as *orbits* or *shells* (see Figure 1-1). Each orbit can accommodate only a limited number of electrons and lies at a fixed distance from the nucleus. Each level must be filled to capacity with electrons before a new level can get started. The orbit closest to the nucleus, which may be referred to as the *first level* or *first shell,* can accommodate up to two electrons. The second level can have eight electrons and the third also can have eight electrons. Higher orbits exist, but anatomy and physiology students only need to know about the first three levels.

Figure 1-1:
Grouping
electrons
into shells
or orbits.

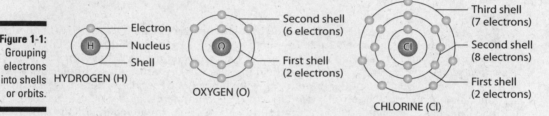

Other key chemistry terms that you need to know as an anatomy and physiology student are

- ✔ **Isotopes:** Atoms of an element that have a different number of neutrons and a different atomic weight than usual. In other words, isotopes are alternate forms of the same chemical element, so they will always have the same number of protons as that element, but a different number of neutrons.

- ✔ **Ions:** Because electrons are relatively far from the atomic nucleus, they are most susceptible to external fields. Atoms that have gained or lost electrons are transformed into ions. Getting an extra electron turns an atom into a negatively charged ion, or *anion,* whereas losing an electron creates a positively charged ion, or *cation.*

To keep anions and cations straight, think like a compulsive dieter: Gaining is negative, and losing is positive.

- ✔ **Acid:** A substance that becomes ionized when placed in solution, producing positively charged hydrogen ions, H^+. An acid is considered a proton donor. (Remember, atoms always have the same number of electrons as protons. Ions are produced when an atom gains or loses electrons.)

- ✔ **Base:** A substance that becomes ionized when placed in solution, producing negatively charged hydroxide ions, $(OH)^-$. Bases are referred to as being more *alkaline* than acids and are known as proton acceptors.

- ✔ **pH (potential of hydrogen):** A mathematical measure on a scale of 0 to 14 of the acidity or alkalinity of a substance. A solution is considered *neutral,* neither acid nor base, if its pH is exactly 7. (Pure water has a pH of 7.) A substance is *basic* if its pH is greater than 7 and *acidic* if its pH is less than 7. Interestingly, skin is considered acidic because it has a pH around 5. Blood, on the other hand, is basic with a pH around 7.4.

Answer these practice questions about atoms and elements:

1. The four key elements that make up most living matter are

a. Carbon, hydrogen, nitrogen, and phosphorus

b. Oxygen, carbon, sulfur, and nitrogen

c. Hydrogen, nitrogen, oxygen, and carbon

d. Nitrogen, potassium, carbon, and oxygen

2. Among the subatomic particles in an atom, the two that have equal weight are

a. Neutrons and electrons

b. Protons and neutrons

c. Positrons and protons

d. Neutrons and positrons

3. For an atom with an atomic number of 19 and an atomic weight of 39, the total number of neutrons is

a. 19

b. 20

c. 39

d. 58

4. Element X has 14 electrons. How many electrons are in its outermost shell?

a. 2

b. 6

c. 14

d. 4

5. A substance that, in water, separates into a large number of hydroxide ions is

a. A weak acid

b. A weak base

c. A strong acid

d. A strong base

6. A hydroxyl, or hydroxide, ion has an oxygen atom

a. Only

b. And an extra electron

c. And a hydrogen atom and an extra electron

d. And a hydrogen atom and one less electron

7.–12. Fill in the blanks to complete the following sentences:

Different isotopes of the same element have the same number of **7.** _____ and **8.** _____ but different numbers of **9.** _____. Isotopes also have different atomic **10.** _____. An atom that gains or loses an electron is called an **11.** _____. If an atom loses an electron, it carries a **12.** _____ charge.

Compounding Chemical Reactions

Atoms tend to arrange themselves in the most stable patterns possible, which means that they have a tendency to complete or fill their outermost electron orbits. They join with other atoms to do just that. The force that holds atoms together in collections known as *molecules* is referred to as a *chemical bond.* There are two main types and some secondary types of chemical bonds:

✔ **Ionic bond:** This chemical bond (shown in Figure 1-2) involves a transfer of an electron, so one atom gains an electron while one atom loses an electron. One of the resulting ions carries a negative charge, and the other ion carries a positive charge. Because opposite charges attract, the atoms bond together to form a molecule.

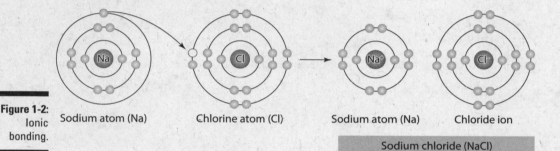

Figure 1-2:
Ionic
bonding.

Sodium atom (Na)   Chlorine atom (Cl)   Sodium atom (Na)   Chloride ion

Sodium chloride (NaCl)

✔ **Covalent bond:** The most common bond in organic molecules, a covalent bond (shown in Figure 1-3) involves the sharing of electrons between two atoms. The pair of shared electrons forms a new orbit that extends around the nuclei of both atoms, producing a molecule. There are two secondary types of covalent bonds that are relevant to biology:

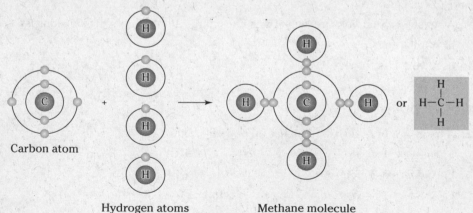

Carbon atom

Figure 1-3:
Covalent
bonding.

Hydrogen atoms   Methane molecule

• **Polar bond:** Two atoms connected by a covalent bond may exert different attractions for the electrons in the bond, producing an unevenly distributed charge. The result is known as a *polar bond,* an intermediate case between ionic and covalent bonding, with one end of the molecule slightly negatively charged and the other end slightly positively charged. Although the resulting molecule is neutral, at close distances the uneven charge distribution can be important. Water is an example of a polar molecule; the oxygen end has a slight positive charge whereas the hydrogen ends are slightly negative. Polarity explains why some substances dissolve readily in water and others do not.

- **Hydrogen bond:** Because they're polarized, two adjacent H_2O (water) molecules can form a linkage known as a _hydrogen bond,_ where a (electronegative) hydrogen atom of one H_2O molecule is electrostatically attracted to the (electropositive) oxygen atom of an adjacent water molecule. Consequently, molecules of water join together transiently in a hydrogen-bonded lattice. Hydrogen bonds have only about $\frac{1}{20}$ the strength of a covalent bond, yet even this force is sufficient to affect the structure of water, producing many of its unique properties, such as high surface tension, specific heat, and heat of vaporization. Hydrogen bonds are important in many life processes, such as in replication and defining the shape of DNA molecules.

A chemical reaction is the result of a process that changes the number, the types, or the arrangement of atoms within a molecule. The substances that go through this process are called the _reactants._ The substances produced by the reaction are called the _products._

Chemical reactions are written in the form of an equation, with an arrow indicating the direction of the reaction. For instance: $A + B \rightarrow AB$. This equation translates to: Atom, ion, or molecule _A_ plus atom, ion, or molecule _B_ yields molecule _AB._

When elements combine through chemical reactions, they form _compounds._ When compounds contain carbon, they're called _organic compounds._ The four families of organic compounds with important biological functions are

- ✔ **Carbohydrates:** These molecules consist of carbon, hydrogen, and oxygen in a ratio of roughly 1:2:1. If a test question involves identifying a compound as a carbohydrate, count the atoms and see if they fit that ratio. Carbohydrates are formed by the chemical reaction process of _concentration,_ or _dehydration synthesis,_ and broken apart by _hydrolysis,_ the cleavage of a chemical by a reaction that adds water. There are several subcategories of carbohydrates:

  - _Monosaccharides,_ also called _monomers_ or _simple sugars,_ are the building blocks of larger carbohydrate molecules and are a source of stored energy (see Figure 1-4). Key monomers include _glucose_ (also known as blood sugar), _fructose,_ and _galactose._ These three have the same numbers of carbon (6), hydrogen (12), and oxygen (6) atoms in each molecule — formally written as $C_6H_{12}O_6$ — but the bonding arrangements are different. Molecules with this kind of relationship are called _isomers._

  - _Disaccharides,_ or _dimers,_ are sugars formed by the bonding of two monosaccharides, including _sucrose_ (table sugar), _lactose,_ and _maltose._

  - _Polysaccharides,_ or _polymers,_ are formed when many monomers bond into long, chain-like molecules. _Glycogen_ is the primary polymer in the body; it breaks down to form glucose, an immediate source of energy for cells.

- ✔ **Lipids:** Commonly known as fats, these molecules contain carbon, hydrogen, and oxygen, and sometimes nitrogen and phosphorous. Insoluble in water because they contain a preponderance of nonpolar bonds, lipid molecules have six times more stored energy than carbohydrate molecules. Upon hydrolysis, however, fats form glycerol and fatty acids. A fatty acid is a long, straight chain of carbon atoms with hydrogen atoms attached (see Figure 1-5). If the carbon chain has its full number of hydrogen atoms, the fatty acid is _saturated_ (examples include butter and lard). If the carbon chain has less than its full number of hydrogen atoms, the fatty acid is _unsaturated_ (examples include margarine and vegetable oils). All fatty acids contain a carboxyl or acid group, –COOH, at the end of the carbon chain. _Phospholipids,_ as the name suggests, contain phosphorus and often nitrogen and form a layer in the cell membrane. _Steroids_ are fat-soluble compounds such as vitamins A or D and hormones that often serve to regulate metabolic processes.

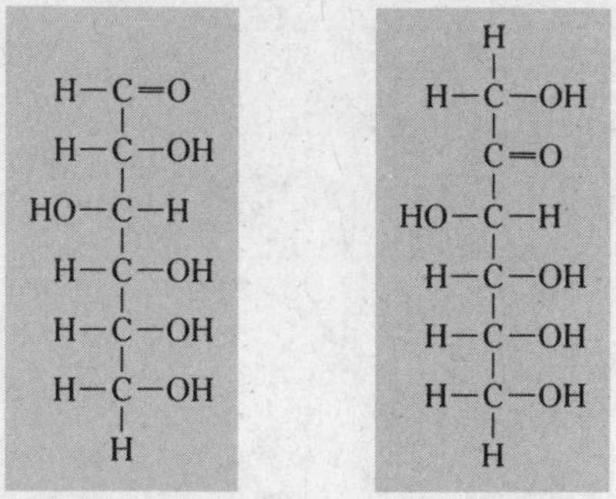

Figure 1-4:
Monosac-
charides.

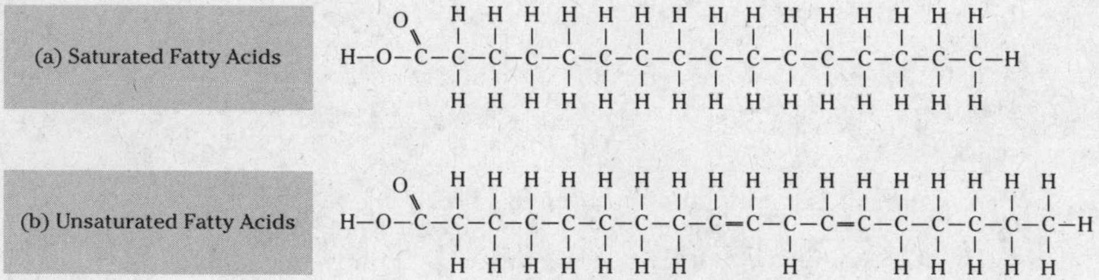

Glucose
$(C_6H_{12}O_6)$

Fractose
$(C_6H_{12}O_6)$

(a) Saturated Fatty Acids

Figure 1-5:
Fatty acids.

(b) Unsaturated Fatty Acids

✔ **Proteins:** Among the largest molecules, proteins can reach molecular weights of some 40 million atomic units. Proteins always contain the four HONC elements — hydrogen, oxygen, nitrogen, and carbon — and sometimes contain phosphorus and sulfur. The human body builds protein molecules using 20 different kinds of smaller molecules called *amino acids* (see Figure 1-6). Each amino acid molecule is composed of an amino group, –NH$_2$, and a carboxyl group, –COOH, with a carbon chain between them. Amino acids link together by peptide bonds to form long molecules called *polypeptides,* which then assemble into proteins. Examples of proteins in the body include *antibodies, hemoglobin* (the red pigment in red blood cells), and *enzymes* (catalysts that accelerate reactions in the body).

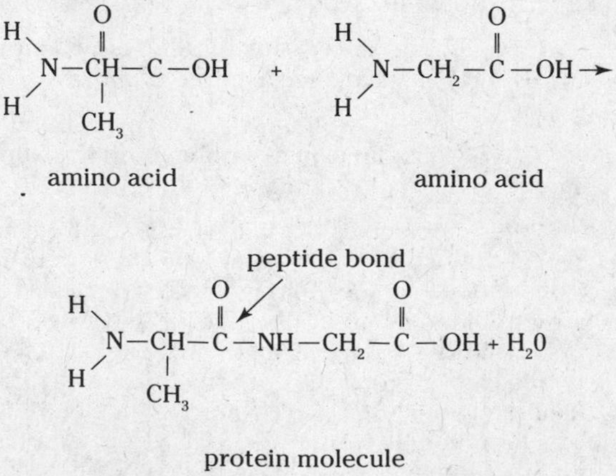

amino acid

amino acid

peptide bond

Figure 1-6:
Amino acids
in a protein
molecule.

protein molecule

✔ **Nucleic acids:** These long molecules, found primarily in the cell's nucleus, act as the body's genetic blueprint. They're comprised of smaller building blocks called *nucleotides.* Each nucleotide, in turn, is composed of a five-carbon sugar (*deoxyribose* or *ribose*), a phosphate group, and a nitrogenous base. The nitrogenous bases in DNA (deoxyribonucleic acid) are *adenine, thymine, cytosine,* and *guanine;* they always pair off A-T and C-G. In RNA (ribonucleic acid), which occurs in a single strand, thymine is replaced by *uracil,* so the nucleotides pair off A-U and C-G. In 1953, James Watson and Francis Crick published their discovery of the three-dimensional structure of DNA — a polymer that looks like a ladder twisted into a coil. They called this structure the *double-stranded helix* (see Figure 1-7).

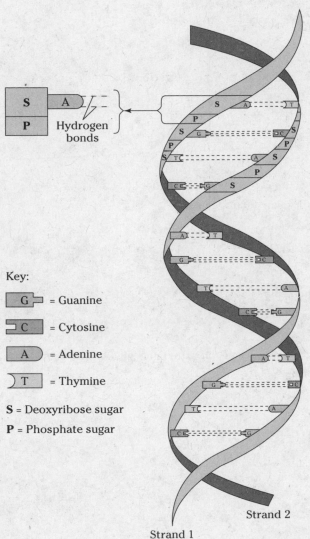

Key:

G = Guanine

C = Cytosine

A = Adenine

T = Thymine

S = Deoxyribose sugar

P = Phosphate sugar

Strand 1

Strand 2

Figure 1-7:
The DNA
double helix.

The following is an example question dealing with chemical reactions:

Q. Oxygen can react with other atoms because it has

   a. Two electrons in its inner orbit

   b. Eight protons

   c. An incomplete outer electron orbit

   d. Eight neutrons

A. The correct answer is an incomplete outer electron orbit. Even if you don't know the first thing about oxygen, remembering that atoms tend toward stability answers this question for you.

13. Bonds formed as a result of sharing one or more electrons between atoms are

   a. Valence bonds

   b. Covalent bonds

   c. Ionic bonds

   d. Electrovalent bonds

14. The formation of chemical bonds is based on the tendency of an atom to

   a. Move protons into vacant electron orbit spaces

   b. Fill its outermost energy level

   c. Radiate excess neutrons

   d. Pick up free protons

15. Which of the following statements is *not* true of DNA?

   a. DNA is found in the nucleus of the cell.

   b. DNA can replicate itself.

   c. DNA contains the nitrogenous bases adenine, thymine, guanine, cytosine, and uracil.

   d. DNA forms a double-helix molecule.

16. Polysaccharides

   a. Can be reduced to fatty acids

   b. Contain nitrogen and phosphorus

   c. Are complex carbohydrates

   d. Contain adenine and uracil

17. Amino acids are the building blocks of

   a. Carbohydrates

   b. Proteins

   c. Lipids

   d. Nucleic acids

Cycling through Life: Metabolism

Metabolism (from the Greek *metabole,* which means "change") is the word for the myriad chemical reactions that happen in the body, particularly as they relate to generating, storing, and expending energy. All metabolic reactions are either *catabolic* or *anabolic. Catabolic reactions* break food down into energy (memory tip: it can be *cata*strophic when things break down). *Anabolic reactions* require the expenditure of energy to build up compounds that the body needs. The chemical alteration of molecules in the cell is referred to as *cellular metabolism. Enzymes* can be used as catalysts, accelerating chemical reactions without being changed by the reactions. The molecules that enzymes react with are called *substrates.*

Adenosine triphosphate (ATP) is a molecule that stores energy in a cell until the cell needs it. As the *tri–* prefix implies, a single molecule of ATP is composed of three phosphate groups attached to a nitrogenous base of adenine. ATP's energy is stored in high-energy bonds that attach the second and third phosphate groups. (The high-energy bond is symbolized by a wavy line.) When a cell needs energy, it removes one or two of those phosphate groups, releasing energy and converting ATP into either the two-phosphate molecule *adenosine diphosphate* (ADP) or the one-phosphate molecule *adenosine monophosphate* (AMP). (You can see ADP and ATP molecules in Figure 1-8.) Later, through additional metabolic reactions, the second and third phosphate groups are reattached to adenosine, reforming an ATP molecule until energy is needed again.

Oxidation-reduction reactions are an important pair of reactions that occur in carbohydrate, lipid, and protein metabolism (see Figure 1-10). When a substance is *oxidized,* it loses electrons and hydrogen ions, removing a hydrogen atom from each molecule. When a substance is *reduced,* it gains electrons and hydrogen ions, adding a hydrogen atom to each molecule. Oxidation and reduction occur together, so whenever one substance is oxidized, another is reduced. The body uses this chemical-reaction pairing to transport energy in a process known as the respiratory chain, or the *electron transport chain.*

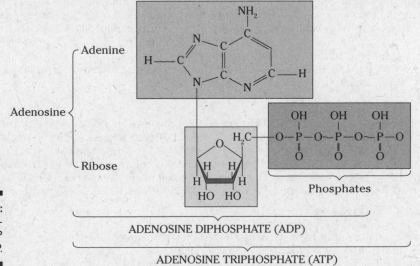

Figure 1-8: The structures of ADP and ATP.

Carbohydrate metabolism involves a series of *cellular respiration* reactions, which are illustrated in Figure 1-9. All food carbohydrates are eventually broken down into glucose; therefore, carbohydrate metabolism is really glucose metabolism. Glucose metabolism produces energy that is then stored in ATP molecules. The oxidation process in which energy is released from molecules, such as glucose, and transferred to other molecules is called cellular respiration. It occurs in every cell in the body and it is the cell's source of energy. The complete oxidation of one molecule of glucose will produce 38 molecules of ATP. It occurs in three stages: *glycolysis,* the *Krebs cycle,* and the *electron transport chain:*

1. **Glycolysis**

    From the Greek *glyco* (sugar) and *lysis* (breakdown), this is the first stage of both *aerobic* (with oxygen) and *anaerobic* (without oxygen) respiration. Using energy from two molecules of ATP and two molecules of NAD$^+$ *(nicotinamide adenine di-nucleotide)*, glycolysis uses a process called *phosphorylation* to convert a molecule of six-carbon glucose — the smallest molecule that the digestive system can produce during the breakdown of a carbohydrate — into two molecules of three-carbon *pyruvic acid* or *pyruvate,* as well as four ATP molecules and two molecules of NADH *(nicotinamide adenine dinucleotide)*. Taking place in the cell's cytoplasm (see Chapter 2), glycolysis doesn't require oxygen to occur. The pyruvate and NADH move into the cell's *mitochondria* (detailed in Chapter 2), where an aerobic (with oxygen) process converts them into ATP.

2. **Krebs cycle**

    Also known as the *tricarboxylic acid cycle* or *citric acid cycle,* this series of energy-producing chemical reactions begins in the mitochondria after pyruvate arrives from glycolysis. Before the Krebs cycle can begin, the pyruvate loses a carbon dioxide group to form *acetyl coenzyme A* (acetyl CoA). Then acetyl CoA combines with a four-carbon molecule (*oxaloacetic acid,* or OAA) to form a six-carbon citric acid molecule that then enters the Krebs cycle. The CoA is released intact to bind with another acetyl group. During the conversion, two carbon atoms are lost as carbon dioxide and energy is released. One ATP molecule is produced each time an acetyl CoA molecule is split. The cycle goes through eight steps, rearranging the atoms of citric acid to produce different intermediate molecules called *keto acids*. The acetic acid is broken apart by carbon (or *decarboxylated*) and oxidized, generating three molecules of NADH, one molecule of FADH2 (flavin adenine dinucleotide), and one molecule of ATP. The energy can be transported to the electron transport chain and used to produce more molecules of ATP. OAA is regenerated to get the next cycle going, and carbon dioxide produced during this cycle is exhaled from the lungs.

3. **Electron transport chain**

    The electron transport chain is a series of energy compounds attached to the inner mitochondrial membrane. The electron molecules in the chain are called *cytochromes*. These electron-transferring proteins contain a heme, or iron, group. Hydrogen from oxidized food sources attaches to coenzymes that in turn combine with molecular oxygen. The energy released during these reactions is used to attach inorganic phosphate groups to ADP and form ATP molecules.

    Pairs of electrons transferred to NAD$^+$ go through the electron transport process and produce three molecules of ATP by oxidative phosphorylation. Pairs of electrons transferred to FAD enter the electron transport after the first phosphorylation and yield only two molecules of ATP. Oxidative phosphorylation is important because it makes energy available in a form the cells can use.

    At the end of the chain, two positively charged hydrogen molecules combine with two electrons and an atom of oxygen to form water. The final molecule to which electrons are passed is oxygen. Electrons are transferred from one molecule to the next, producing ATP molecules.

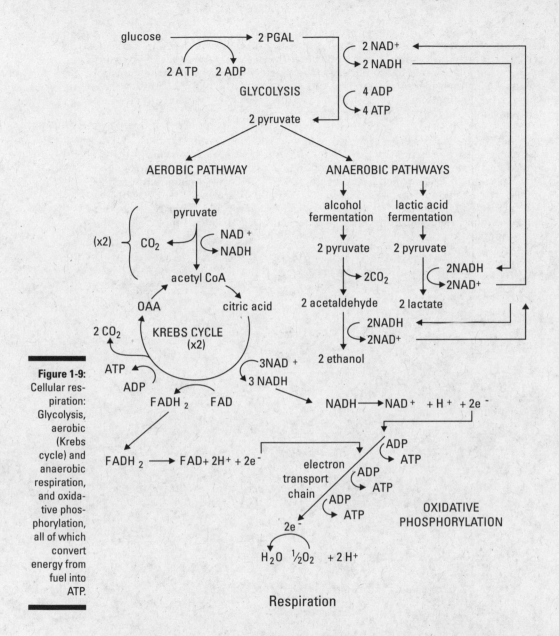

Figure 1-9: Cellular respiration: Glycolysis, aerobic (Krebs cycle) and anaerobic respiration, and oxidative phosphorylation, all of which convert energy from fuel into ATP.

Respiration

Lipid metabolism only requires portions of the processes involved in carbohydrate metabolism. Lipids contain about 99 percent of the body's stored energy and can be digested at mealtime, but as people who complain about fats going "straight to their hips" can attest, lipids are more inclined to be stored in *adipose tissue* — the stuff generally identified with body fat. When the body is ready to metabolize lipids, a series of catabolic reactions breaks apart two carbon atoms from the end of a fatty acid chain to form acetyl CoA, which then enters the Krebs cycle to produce ATP. Those reactions continue to strip two carbon atoms at a time until the entire fatty acid chain is converted into acetyl CoA.

Protein metabolism focuses on producing the amino acids needed for synthesis of protein molecules within the body. But in addition to the energy released into the electron transport chain during protein metabolism, the process also produces byproducts, such as ammonia and keto acid. Energy is released entering the electron transport chain. The liver converts the ammonia into urea, which the blood carries to the kidneys for elimination. The keto acid enters the Krebs cycle and is converted into pyruvic acids to produce ATP.

One last thing: That severe soreness and fatigue you feel in your muscles after strenuous exercise is the result of lactic acid buildup during *anaerobic respiration*. Glycolysis continues because it doesn't need oxygen to take place. But glycolysis does need a steady supply of NAD⁺, which usually comes from the oxygen-dependent electron transport chain converting NADH back into NAD⁺. In its absence, the body begins a process called *lactic acid fermentation,* in which one molecule of pyruvate combines with one molecule of NADH to produce a molecule of NAD⁺ plus a molecule of the toxic byproduct lactic acid.

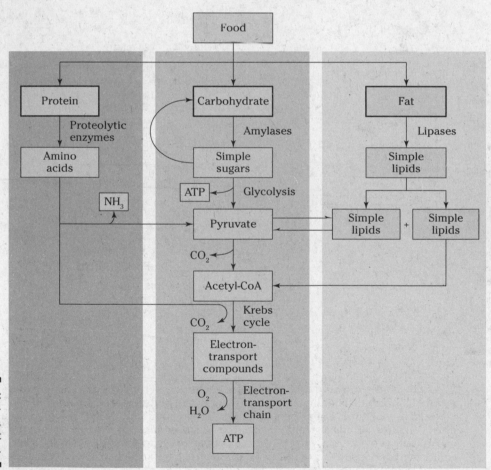

Figure 1-10:
Protein, carbohydrate, and fat metabolism.

Check out an example question on metabolism:

EXAMPLE

0. Cells obtain ATP by converting the energy in

  a. Carbohydrates

  b. Proteins

  c. Lipids

  d. All of these

A. The correct answer is all of these. While it's true that carbohydrates provide the most immediately available energy, proteins and lipids also contribute to the production of ATP.

18. A molecule of glucose is broken down to pyruvic acid by

  a. Glycolysis

  b. The Krebs cycle

  c. The electron transport chain

  d. Oxidative phosphorylation

19. Pyruvic acid enters a mitochondrion and is converted into

  a. Glucose

  b. Acetyl CoA

  c. Water

  d. Protein

20. A molecule of glucose can be converted into how many ATP molecules?

  a. 2

  b. 3

  c. 38

  d. 45

21. The part of metabolism that involves creating compounds the body needs is called

  a. A catabolic reaction

  b. Cellular respiration

  c. An anabolic reaction

  d. Oxidation

22. Metabolic processes that don't require oxygen are called

  a. Anaerobic

  b. Aerobic

  c. Fermentation

  d. Carbon dioxination

23. Which two respiration processes take place in the cell's mitochondria?

    a. Glycolysis and the Krebs cycle

    b. Glycolysis and the electron transport chain

    c. The Krebs cycle and the electron transport chain

    d. The Krebs cycle and anaerobic respiration

24. Coal is to electricity as glucose is to

    a. ATP

    b. Pyruvate

    c. Hydrogen

    d. Glycolysis

25. The primary products of protein metabolism are

    a. ATP molecules

    b. Amino acids

    c. Lipids

    d. Carbon dioxide molecules

26. Fats are metabolized primarily during

    a. Glycolysis

    b. Lactic acid fermentation

    c. Exercise

    d. The Krebs cycle

Answers to Questions on Life's Chemistry

The following are answers to the practice questions presented in this chapter.

1. The four key elements that make up most living matter are **c. hydrogen, nitrogen, oxygen, and carbon.** We arranged them so that they spell HNOC instead of HONC, but you get the idea, right?

2. Among the subatomic particles in an atom, the two that have equal weight are **b. protons and neutrons.** That's why you add them together to determine atomic weight, or mass.

3. For an atom with an atomic number of 19 and an atomic weight of 39, the total number of neutrons is **b. 20.** The atomic number of 19 is the same as the number of protons. The atomic weight of 39 tells you the number of protons plus the number of neutrons: 39 – 19 = 20.

4. Element X has 14 electrons. How many electrons are in its outermost shell? **d. 4.** The first orbit has the maximum two electrons, and the second orbit has the maximum eight electrons. That makes ten electrons in the first two orbits, leaving only four for the third, outermost orbit.

5. A substance that, in water, separates into a large number of hydroxide ions is **d. a strong base.** The more hydroxide ions there are, the stronger the base is.

6. A hydroxyl, or hydroxide, ion has an oxygen atom **c. and a hydrogen atom and an extra electron.** The first few letters of the word "hydroxide" are a dead giveaway that there's a hydrogen atom in there; plus hydroxide ions are negatively charged, which calls for that extra electron.

7.–12. Different isotopes of the same element have the same number of **7. electrons/protons** and **8. protons/electrons** but different numbers of **9. neutrons.** Isotopes also have different atomic **10. weights.** An atom that gains or loses an electron is called an **11. ion.** If an atom loses an electron, it carries a **12. positive** charge.

13. Bonds formed as a result of sharing one or more electrons between atoms are **b. covalent bonds.** If the atoms had gained or lost electrons, it would be an ionic bond, but here they're sharing — valiantly cohabiting, if you will.

14. The formation of chemical bonds is based on the tendency of an atom to **b. fill its outermost energy level.** This is true whether an atom fills its outer shell by sharing, gaining, or losing electrons.

15. Which of the following statements is *not* true of DNA? **c. DNA contains the nitrogenous bases adenine, thymine, guanine, cytosine, and uracil.** This statement is false because only RNA contains uracil.

16. Polysaccharides **c. are complex carbohydrates.** The root *poly–* means "many," which you can interpret as "complex." The root *mono–* means "one," which you can interpret as "simple."

17. Amino acids are the building blocks of **b. proteins.** Being such large molecules, proteins need to be built from complex molecules to begin with.

18. A molecule of glucose is broken down to pyruvic acid by **a. glycolysis.** Remember that glucose must become pyruvic acid before it enters the Krebs cycle.

19. Pyruvic acid enters a mitochondrion and is converted into **b. acetyl CoA.** Don't forget that the Krebs cycle, during which pyruvate is broken down, occurs in the mitochondrion.

20 A molecule of glucose can be converted into how many ATP molecules? **c. 38.** Two net molecules of ATP come from glycolysis, two molecules come from the Krebs cycle, and the electron transport chain churns out 34.

21 The part of metabolism that involves creating compounds the body needs is called **c. an anabolic reaction.** Breaking things down is a catabolic reaction, but building them up is anabolic.

22 Metabolic processes that don't require oxygen are called **a. anaerobic.** Recall that during aerobic exercise, you're trying to circulate oxygen to your muscles. So anaerobic is the opposite.

23 Which two respiration processes take place in the cell's mitochondria? **c. The Krebs cycle and the electron transport chain.** The other answers are incorrect because glycolysis takes place in the cytoplasm, and anaerobic respiration isn't one of the three cellular respiration processes.

24 Coal is to electricity as glucose is to **a. ATP.** Just as you can't power a lamp with a lump of coal, cells can't use glucose directly. You need to turn the coal into electricity, and cells need to turn the glucose into ATP.

25 The primary products of protein metabolism are **b. amino acids.** Although some ATP comes from metabolizing proteins, the body primarily needs to get amino acids from any protein that's consumed.

26 Fats are metabolized primarily during **d. the Krebs cycle.** That's the only process that can use the acetyl CoA supplied by lipids.

Chapter 2

The Cell: Life's Basic Building Block

Cytology, from the Greek word *cyto,* which means "cell," is the study of cells. Every living thing has cells, but not all living things have the same kinds of cells. *Eukaryotes* like humans (and all other organisms besides bacteria and viruses) have *eukaryotic cells,* each of which has a defined *nucleus* that controls and directs the cell's activities, and *cytosol,* fluid material found in the gel-like *cytoplasm* that fills most of the cell. Plant cells have fibrous cell walls; animal cells do not, making do instead with a semipermeable *cell membrane,* which sometimes is called a *plasma membrane* or the *plasmalemma.* Because human cells don't have cell walls, they look like gel-filled sacs with nuclei and tiny parts called *organelles* nestled inside when viewed through an electron microscope.

In this chapter, we help you sort out what makes up a cell, what all those tiny parts do, and how cells act as protein-manufacturing plants to support life's activities. We then take a quick look at an individual cell's life cycle.

Gaining Admission: The Cell Membrane

Think of it as a gatekeeper, guardian, or border guard. Despite being only 6 to 10 nanometers thick and visible only through an electron microscope, the cell membrane keeps the cell's cytoplasm in place and lets only select materials enter and depart the cell as needed. This semipermeability, or *selective permeability,* is a result of a double layer (bilayer) of *phospholipid* molecules interspersed with protein molecules. The outer surface of each layer is made up of tightly packed *hydrophilic* (or water-loving) *polar heads.* Inside, between the two layers, you find *hydrophobic* (or water-fearing) *nonpolar tails* consisting of fatty acid chains. Cholesterol molecules between the phosphate layers give the otherwise elastic membrane stability and make it less permeable to water-soluble substances. Both cytoplasm and the *matrix,* the material in which cells lie, are primarily water. The polar heads electrostatically attract polarized water molecules while the nonpolar tails lie between the layers, shielded from water and creating a dry middle layer. The membrane's interior is made up of oily fatty acid molecules that are electrostatically symmetric, or *nonpolarized.* Lipid-soluble molecules can pass through this layer, but water-soluble molecules such as amino acids, sugars, and proteins cannot. Because phospholipids have both polar and nonpolar regions, they're also called *amphipathic molecules.*

The cell membrane is designed to hold the cell together and to isolate it as a distinct functional unit of protoplasm. Although it can spontaneously repair minor tears, severe damage to the membrane will cause the cell to disintegrate. The membrane is picky about which molecules it lets in or out. It allows movement across its barrier by *diffusion, osmosis,* or *active transport* as follows:

✔ **Diffusion:** This is a spontaneous spreading, or migration, of molecules or other particles from an area of higher concentration to an area of lower concentration until equilibrium occurs. When equilibrium is reached, diffusion continues, but the flow is equal in both directions. Diffusion is a natural phenomenon that behaves in much the same way as *Brownian motion;* both phenomena are based on the fact that all molecules possess kinetic energy. They move randomly at high speeds, colliding with one another, changing directions, and moving away from areas of greatest concentration to areas of lower concentration. The rate of movement depends on the size and temperature of the molecule; the smaller and warmer the molecule is, the faster it moves.

Diffusion is one form of *passive transport* that doesn't require the expenditure of cellular energy. A molecule can diffuse passively through the cell membrane if it's lipid-soluble, uncharged, and very small, or if it can be assisted by a carrier molecule. The unassisted diffusion of very small or lipid-soluble particles is called *simple diffusion.* The assisted process is known as *facilitated diffusion.* The cell membrane allows nonpolar molecules (those that don't readily bond with water) to flow from an area where they're highly concentrated to an area where they're less concentrated. Embedded with the hydrophilic heads in the outer layer are protein molecules called *channel proteins* that create diffusion-friendly openings for the molecules to diffuse through.

✔ **Osmosis:** This form of passive transport is similar to diffusion and involves a solvent moving through a selectively permeable or semipermeable membrane from an area of higher concentration to an area of lower concentration. Solutions are composed of two parts: a *solvent* and a *solute.* The solvent is the liquid in which a substance is dissolved; water is called the *universal solvent* because more materials dissolve in it than in any other liquid. A *solute* is the substance dissolved in the solvent. Typically, a cell contains a roughly 1 percent saline solution — in other words, 1 percent salt (solute) and 99 percent water (solvent). Water is a polar molecule that will not pass through the lipid bilayer; however, it is small enough to move through the pores of most cell membranes. Osmosis occurs when there's a difference in molecular concentration of water on the two sides of the membrane. The membrane allows the solvent (water) to move through but keeps out the particles dissolved in the water.

Transport by osmosis is affected by the concentration of solute (the number of particles) in the water. One molecule or one ion of solute displaces one molecule of water. *Osmolarity* is the term used to describe the concentration of solute particles per liter. As water diffuses into a cell, hydrostatic pressure builds within the cell. Eventually, the pressure within the cell becomes equal to, and is balanced by, the osmotic pressure outside.

- An *isotonic* solution has the same concentration of solute and solvent as found inside a cell, so a cell placed in isotonic solution — typically 1 percent saline solution for humans — experiences equal flow of water into and out of the cell, maintaining equilibrium.

- A *hypotonic* solution has less solute and higher water potential than inside the cell. An example is 100 percent distilled water, which has less solute than what's inside the cell. Therefore, if a human cell is placed in a hypotonic solution, molecules diffuse down the concentration gradient until the cell's membrane bursts.

- A *hypertonic* solution has more solute and lower water potential than inside the cell. So the membrane of a human cell placed in 10 percent saline solution (10 percent salt and 90 percent water) would let water flow out of the cell (from higher concentration inside to lower concentration outside), therefore shrinking it.

✔ **Active transport:** This movement occurs across a semipermeable membrane against the normal concentration gradient, moving from the area of lower concentration to the area of higher concentration and requiring an expenditure of energy released from an ATP molecule (as discussed in Chapter 1). Embedded with the hydrophilic heads in the outer layer of the membrane are protein molecules able to detect and move compounds through the membrane. These *carrier* or *transport* proteins interact with the *passenger* molecules and use the ATP-supplied energy to move them against the gradient. The carrier molecules combine with the transport molecules — most importantly amino acids and ions — to pump them against their concentration gradients.

Active transport lets cells obtain nutrients that can't pass through the membrane by other means. In addition, there are secondary active transport processes that are similar to diffusion but instead use imbalances in electrostatic forces to move molecules across the membrane.

1.–3. Fill in the blanks to complete the following sentences:

The lipid bilayer structure of the cell membrane is made possible because phospholipid molecules contain two distinct regions: The **1.** _____ region is attracted to water, and the **2.** _____ region is repelled by water. Because it has both polar and non-polar regions, a phospholipid is classified as a(n) **3.** _____ molecule.

4. The movement of water molecules through a semipermeable membrane is known as

   a. Diffusion

   b. Filtration

   c. Osmosis

   d. Active transport

5. A solution having a greater concentration of water than exists in the cell is said to be

   a. Hypertonic

   b. Hypotonic

   c. Isotonic

   d. Heterotonic

6. Injecting a large quantity of distilled water into a human's veins would cause many red blood cells to

   a. Swell and burst

   b. Shrink

   c. Carry more oxygen

   d. Aggregate

7. The cell membrane does *not* function

   a. In selective transport of materials into and out of the cell

   b. As a barrier protecting the cell

   c. In the production of energy

   d. As containment for the cytoplasm

Aiming for the Nucleus

The cell nucleus is the largest cellular organelle and the first to be discovered by scientists. On average, it accounts for about 10 percent of the total volume of the cell, and it holds a complete set of genes.

The outermost part of this organelle is the *nuclear envelope,* which is composed of a double-membrane barrier, each membrane of which is made up of a phospholipid bilayer. Between the two membranes is a fluid-filled space called the *perinuclear cisterna.* The two layers fuse to form a selectively permeable barrier, but large *pores* allow relatively free movement of molecules and ions, including large protein molecules. Intermediate filaments lining the surface of the nuclear envelope make up the *nuclear lamina,* which functions in the disassembly and reassembly of the nuclear membrane during mitosis and binds the membrane to the endoplasmic reticulum. The nucleus also contains *nucleoplasm,* a clear viscous material that forms the matrix in which the organelles of the nucleus are embedded.

DNA is packaged inside the nucleus in structures called *chromatin,* or *chromatin networks.* During cell division, the chromatin contracts, forming *chromosomes.* Chromosomes contain a DNA molecule encoded with the genetic information needed to direct the cell's activities. The most prominent subnuclear body is the *nucleolus,* a small spherical body that stores RNA molecules and produces *ribosomes,* which are exported to the cytoplasm where they translate *messenger RNA* (mRNA).

The following is an example question about the nucleus:

EXAMPLE

Q. The only cellular organelle found within the nucleus is called a(n)

  a. Lamina

  b. Envelope

  c. Nucleolus

  d. Chromosome

A. The correct answer is nucleolus. The other options aren't considered organelles.

8. The fluid-filled space within the nuclear envelope is called the

  a. Perinuclear cisterna

  b. Ribosome

  c. Mitochondrion

  d. Golgi appartus

9. DNA is packaged within

  a. Chromatins

  b. Chromosomes

  c. Ribosomes

  d. Genes

10. The nucleolus

   a. Packages DNA

   b. Enables large molecule transport

   c. Forms a membrane around the nucleus

   d. Assembles ribosomes

Looking Inside: Organelles and Their Functions

Molecules that pass muster with the cell membrane enter the *cytoplasm,* a mixture of macromolecules such as proteins and RNA and small organic molecules such as glucose, ions, and water. Because of the various materials in the cytoplasm, it's a *colloid,* or mixture of phases, that alternates from a *sol* (a liquid colloid with solid suspended in it) to a *gel* (a colloid in which the dispersed phase combines with the medium to form a semisolid material). The fluid part of the cytoplasm, called the *cytosol,* has a differing consistency based on changes in temperature, molecular concentrations, pH, pressure, and agitation.

Within the cytoplasm lies a network of fibrous proteins collectively referred to as the *cytoskeleton.* It's not rigid or permanent but changing and shifting according to the activity of the cell. The cytoskeleton maintains the cell's shape, enables it to move, anchors its organelles, and directs the flow of the cytoplasm. The fibrous proteins that make up the cytoskeleton include the following:

   ✔ *Microfilaments,* rodlike structures about 5 to 8 nanometers wide that consist of a stacked protein called *actin,* the most abundant protein in eukaryotic cells. They provide structural support and have a role in cell and organelle movement as well as in cell division.

   ✔ *Intermediate filaments*, the strongest and most stable part of the cytoskeleton. They average about 10 nanometers wide and consist of interlocking proteins, including *keratin,* that chiefly are involved in maintaining cell integrity and resisting pulling forces on the cell.

   ✔ Hollow *microtubules* about 25 nanometers in diameter that are made of the protein *tubulin* and grow with one end embedded in the *centrosome* near the cell's nucleus. Like microfilaments, these components of cilia, flagella, and centrioles provide structural support and have a role in cell and organelle movement as well as in cell division.

Organelles, literally translated as "little organs," are nestled inside the cytoplasm (except for the two organelles that move, *cilia* and *flagellum,* which are found on the cell's exterior). Each organelle has different responsibilities for producing materials used elsewhere in the cell or body. Here are the key organelles and what they do:

   ✔ **Centrosome:** Microtubules sprout from this structure, which is located next to the nucleus and is composed of two *centrioles* — arrays of microtubules — that function in separating genetic material during cell division.

   ✔ **Cilia:** These are short, hair-like cytoplasmic projections on the external surface of the cell. In multicellular animals, including humans, cilia move materials over the surface of the cell. In some single-celled organisms, they're used for locomotion.

- **Endoplasmic reticulum (ER):** This organelle makes direct contact with the cell nucleus and functions in the transport of materials such as proteins and RNA molecules. Composed of membrane-bound canals and cavities that extend from the nuclear membrane to the cell membrane, the ER is the site of lipid and protein synthesis. The two types of ER are *rough,* which is dotted with ribosomes on the outer surface; and *smooth,* which has no ribosomes on the surface.

- **Flagellum:** This whip-like cytoplasmic projection lies on the cell's exterior surface. Found in humans primarily on sperm cells, it's used for locomotion.

- **Golgi apparatus (or body):** This organelle consists of a stack of flattened sacs with membranes that connect with those of the endoplasmic reticulum. Located near the nucleus, it functions in the storage, modification, and packaging of proteins for secretion to various destinations within the cell.

- **Lysosome:** A tiny, membranous sac containing acids and digestive enzymes, the lysosome breaks down large food molecules such as proteins, carbohydrates, and nucleic acids into materials that the cell can use. It destroys foreign particles in the cell and helps to remove nonfunctioning structures from the cell.

- **Mitochondrion:** Called the powerhouse of the cell, this rod-shaped organelle consists of two membranes — a smooth outer membrane, and an invaginated (folded inward) inner membrane that divides the organelle into compartments. The inward-folding crevices of the inner membrane are called *cristae.* The mitochondrion provides critical functions in cell respiration, including oxidizing (breaking down) food molecules and releasing energy that is stored in ATP molecules in the mitochondrion. This energy is used to accelerate chemical reactions in the cell, which we cover in Chapter 1.

- **Ribosomes:** These roughly 25-nanometer structures may be found along the endoplasmic reticulum or floating free in the cytoplasm. Composed of 60 percent RNA and 40 percent protein, they translate the genetic information on RNA molecules to *synthesize,* or produce, a protein molecule.

- **Vacuoles:** More commonly found in plant cells, these open spaces in the cytoplasm sometimes carry materials to the cell membrane for discharge to the outside of the cell. In animal cells, food vacuoles are membranous sacs formed when food masses are pinched-off from the cell membrane and passed into the cytoplasm of the cell. This process, called *endocytosis* (from the Greek words meaning "within the cell"), requires energy to move large masses of material into the cell. Vacuoles also help to remove structural debris, isolate harmful materials, and export unwanted substances from the cell.

Answer these practice questions about cell organelles:

11. This cigar-shaped organelle produces energy through aerobic respiration.

a. Golgi apparatus

b. Mitochondrion

c. Lysosome

d. Endoplasmic reticulum

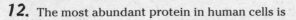

12. The most abundant protein in human cells is
 a. Actin
 b. Tubulin
 c. Albumen
 d. Cytoplasm

13. Which organelle gets to take out the cellular trash?
 a. Golgi apparatus
 b. Ribosome
 c. Vacuole
 d. Endoplasmic reticulum

14. The very small organelle responsible for protein synthesis (making proteins) is the
 a. Ribosome
 b. Lysosome
 c. Centriole
 d. Vesicle

15. Which organelle has ribosomes attached to it?
 a. Smooth (agranular) endoplasmic reticulum
 b. Golgi apparatus
 c. Rough (granular) endoplasmic reticulum
 d. Nucleus

16. Which organelle contains secretory materials?
 a. Golgi apparatus
 b. Ribosome
 c. Lysosome
 d. Endoplasmic reticulum

17. Which of the following can change the consistency of cytoplasm?
 a. Changes in acidity or alkalinity
 b. Temperature
 c. Pressure
 d. All of the above

18. Structures found inside the nucleus include the
 a. Mitochondria
 b. Lysosomes
 c. Chromatin network
 d. Ribosomes

19.–32. Use the terms that follow to identify the cell structures and organelles shown in Figure 2-1.

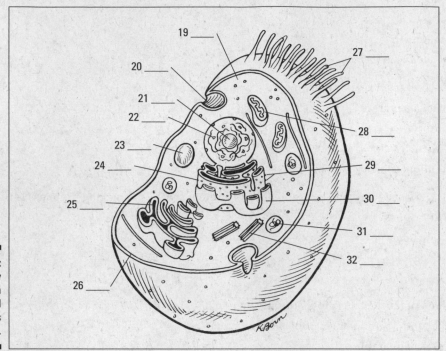

Figure 2-1:
A cutaway
view of an
animal cell
and its
organelles.

a. Centrioles

b. Cilia

c. Cytoplasm

d. Golgi apparatus

e. Lysosome

f. Mitochondrion

g. Nucleolus

h. Nucleus

i. Plasma (cell) membrane

j. Ribosomes

k. Rough endoplasmic reticulum

l. Smooth endoplasmic reticulum

m. Vacuoles

n. Vesicle formation

33.–37. Match the organelles with their descriptions.

33. _____ Mitochondrion

34. _____ Nucleolus

35. _____ Flagellum

36. _____ Cytoplasm

37. _____ Lysosomes

a. Long, whip-like organelle for locomotion

b. Fluid-like interior of the cell that may become a semisolid, or colloid

c. Membranous sacs containing digestive enzymes

d. Powerhouse of the cell

e. Stores RNA in the nucleus

Putting Together New Proteins

Proteins are essential building blocks for all living systems, which helps explain why the word is derived from the Greek term *proteios,* meaning "holding first place." Cells use proteins to perform a variety of functions, including providing structural support and catalyzing reactions. Cells synthesize proteins through a systematic procedure that begins in the nucleus when the gene code for a certain protein is *transcribed* from the cell's DNA into *messenger RNA,* or *mRNA.* The mRNA moves through nuclear pores to the rough endoplasmic reticulum (ER), where ribosomes *translate* the message one *codon* of three nucleotides, or *base pairs,* at a time. The ribosome uses *transfer RNA,* or *tRNA,* to fetch each required amino acid and then link them together through peptide bonds, also known as amide bonds, to form proteins (see Figure 2-2 for details).

Don't let the labels confuse you. Proteins are chains of amino acids (usually very long chains of at least 100 acids). Enzymes, used to catalyze reactions, also are chains of amino acids and therefore also are categorized as proteins. *Polypeptides,* or simply *peptides,* are shorter chains of amino acids used to bond larger protein molecules, but they also can be regarded as proteins. Both antibodies and hormones also are proteins, along with almost everything else in the body — hair, muscle, cartilage, and so on. Even the four basic blood types — A, B, AB, and O — are differentiated by the proteins found in each.

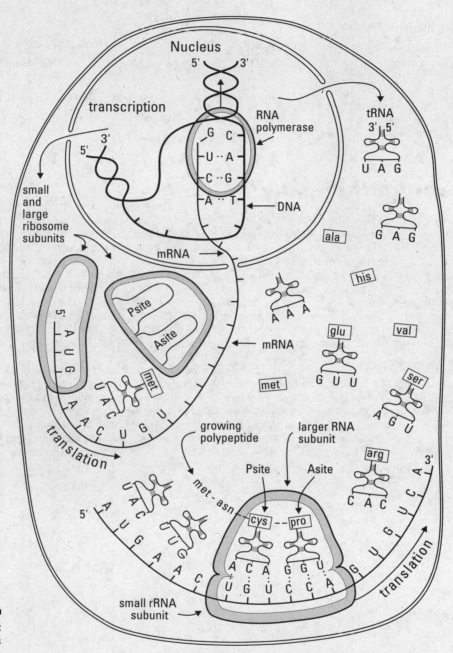

Figure 2-2:
The process
of protein
synthesis.

Protein Synthesis

38.–44. Fill in the blanks to complete the following sentences:

Protein synthesis begins in the cell's **38.** _____ when the gene for a certain protein is **39.** _____ into messenger RNA, or mRNA, which then moves on to the **40.** _____. There, ribosomes **41.** _____ three base pairs at a time, forming a series also referred to as a **42.** _____. Molecules called **43.** _____ then collect each amino acid needed so that the ribosomes can link them together through **44.** _____ bonds.

45. The word "protein" can refer to

   a. Hormones

   b. Enzymes

   c. Antibodies

   d. All of the above

46. Which of the following comes first in the protein-synthesis process?

   a. Transfer RNA

   b. Transcription

   c. Peptide bonds

   d. Translation

47. tRNA is used to gather

   a. Blood cells

   b. Amino acids

   c. Protein molecules

   d. DNA

48. A codon is a sequence of three

   a. Nucleotides

   b. Amino acids

   c. Ribosomes

   d. Base pieces

Cycling Along: Grow, Rest, Divide, Die

The *cell life cycle,* usually referred to simply as the *cell cycle* or the *CDC (cell division cycle),* extends from the beginning of one cell division to the beginning of the next division. The human body produces new cells every day to replace those that are damaged or worn out.

The cell cycle is divided into two distinct phases:

> ✔ **Interphase:** Sometimes also called the resting stage, that label is a misnomer because the cell is actively growing and carrying out its normal metabolic functions as well as preparing for cell division.
>
> ✔ **Mitosis:** The period of cell division that produces new cells. (We cover this phase in detail in Chapter 3.)

New cells are produced for growth and to replace the billions of cells that stop functioning in the adult human body every day. Some cells, like blood and skin cells, are continually dividing because they have very short life cycles, sometimes only hours. Other cells, such as specialized muscle cells and certain nerve cells, may never divide at all.

49. Human cells can live

  a. A few hours

  b. A few days

  c. Indefinitely

  d. All of the above

50. The cell cycle is measured

  a. By the number of times a cell divides

  b. From the beginning to the end of one cell division

  c. From the beginning of one cell division to the beginning of the next

  d. Slowly, over time

Answers to Questions on the Cell

The following are answers to the practice questions presented in this chapter.

1–3 The lipid bilayer structure of the cell membrane is made possible because phospholipid molecules contain two distinct regions: The **1. hydrophilic** region is attracted to water, and the **2. hydrophobic** region is repelled by water. Because it has both polar and nonpolar regions, a phospholipid is classified as a(n) **3. amphipathic** molecule.

4 The movement of water molecules through a semipermeable membrane is known as **c. osmosis.** Why not diffusion? Because diffusion has to do with the passive transport of substances *other than* water.

5 A solution having a greater concentration of water than exists in the cell is said to be **b. hypotonic.**

The prefix *hypo* refers to under or below normal. The prefix *hyper* refers to excess, or above normal. Someone who has been out in the cold too long suffers hypothermia — literally insufficient heat. So a solution, or tonic, with very few particles would be hypotonic.

6 Injecting a large quantity of distilled water into a human's veins would cause many red blood cells to **a. swell and burst.** With more water outside the cell than inside, the membrane would allow osmosis to continue past the breaking point.

7 The cell membrane does *not* function **c. in the production of energy.** Sometimes it may use energy in the form of ATP, but the cell membrane isn't involved directly in the production of energy.

8 The only organelle with direct links to the nucleus is the **a. endoplasmic reticulum.** The nuclear lamina, or intermediate filaments, link the ER with the nucleus.

9 DNA is packaged within **b. chromosomes.** None of the other answer options contains a full package of DNA.

10 The nucleolus **d. assembles ribosomes.** It's not just a coincidence that the nucleolus sits at the heart of the genetic powerhouse.

11 This cigar-shaped organelle produces energy through aerobic respiration. **b. Mitochondrion.** None of the other options are involved in cellular respiration or energy production

12 The most abundant protein in human cells is **a. actin.** It makes sense that the protein making up much of the cytoskeleton is the most abundant because the cytoskeleton accounts for up to 50 percent of the cell's volume.

13 Which organelle gets to take out the cellular trash? **c. Vacuole**

When it's time to clean house, you pull out the *vacuum*. Cells pull out the *vacuoles*.

14 The very small organelle responsible for protein synthesis (making proteins) is the **a. ribosome.**

When you think of a big protein-laden meal, you think of ribs. Ribs. Ribosome. Protein synthesis. Get it?

15 Which organelle has ribosomes attached to it? **c. Rough (granular) endoplasmic reticulum.** This is a good example of using previous questions to answer later ones.

16 Which organelle contains secretory materials? **a. Golgi apparatus.** The correct answer can't be the lysosome because that's already a vesicle, and it can't be the ribosome because you already know that handles proteins.

17 Which of the following can change the consistency of cytoplasm? **d. All of the above.** In addition, molecular concentration and agitation also can change cytoplasmic consistency.

18 Structures found inside the nucleus include the **c. chromatin network.** Don't confuse the organelles in the cytoplasm with the organelles in the nucleus.

19 – 32 Following is how Figure 2-1, the cutaway view of the cell and its organelles, should be labeled.

> 19. **c. Cytoplasm**; 20. **n. Vesicle formation**; 21. **g. Nucleolus**; 22. **h. Nucleus**; 23. **m. Vacuoles**; 24. **k. Rough endoplasmic reticulum**; 25. **d. Golgi apparatus**; 26. **i. Plasma (cell) membrane**; 27. **b. Cilia**; 28. **f. Mitochondrion**; 29. **j. Ribosomes**; 30. **l. Smooth endoplasmic reticulum**; 31. **e. Lysosome**; 32. **a. Centriole**

33 Mitochondrion: **d. Powerhouse of the cell**

34 Nucleolus: **e. Stores RNA in the nucleus**

35 Flagellum: **a. Long, whip-like organelle for locomotion**

36 Cytoplasm: **b. Fluid-like interior of the cell that may become a semisolid, or colloid**

37 Lysosomes: **c. Membranous sacs containing digestive enzymes**

38 – 44 Protein synthesis begins in the cell's **38. nucleus** when the gene for a certain protein is **39. transcribed** into messenger RNA, or mRNA, which then moves on to the **40. rough endoplasmic reticulum.** There, ribosomes **41. translate** three base pairs at a time, forming a series also referred to as a **42. codon.** Molecules called **43. tRNA** then collect each amino acid needed so that the ribosomes can link them together through **44. peptide** bonds.

45 The word "protein" can refer to **d. all of the above.** Proteins come in myriad shapes, sizes, and functions.

46 Which of the following comes first in the protein synthesis process? **b. Transcription.** Remember that you have to transcribe before you can translate.

47 tRNA is used to gather **b. amino acids.** No amino acids = no protein. Simple as that.

48 A codon is a sequence of three **a. nucleotides.** It takes three to round up a single amino acid.

49 Human cells can live **d. all of the above.** Cell life cycles can vary widely.

50 The cell cycle is measured **c. from the beginning of one cell division to the beginning of the next.**

Chapter 3

Divide and Conquer: Cellular Mitosis

Ever had so many places to be that you wished you could just divide yourself in two? Your cells already do that. Cell division is how one "mother" cell becomes two identical twin "daughter" cells. Cell division takes place for several reasons:

✔ **Growth:** Multicellular organisms, humans included, each start out as a single cell — the fertilized egg. That one cell divides (and divides and divides), eventually becoming an entire complex being.

✔ **Injury repair:** Uninjured cells in the areas surrounding damaged tissue divide to replace those that have been destroyed.

✔ **Replacement:** Cells eventually wear out and cease to function. Their younger, more functional neighbors divide to take up the slack.

✔ **Asexual reproduction:** No, human cells don't do this. Only single-celled organisms do.

Cell division occurs over the course of two processes: *mitosis,* which is when the chromosomes within the cell's nucleus duplicate to form two daughter nuclei; and *cytokinesis,* which takes place when the cell's cytoplasm divides to surround the two newly formed nuclei. Although cell division breaks down into several stages, there are no pauses from one step to another. Cell division as a whole is called mitosis because most of the changes occur during that process. Cytokinesis doesn't start until later. But mitosis and cytokinesis do end together.

Keep in mind: Cells are living things, so they mature, reproduce, and die. In this chapter, we review the cell cycle (as mitosis also is known), and you get plenty of practice figuring out what happens when and why.

The Mitotic Process

It may look like cells are living out their useful lives simply doing whatever specialized jobs they do best, but in truth mitosis is a continuous process. When the cell isn't actively splitting itself in two, it's actively preparing to do so. DNA and centrioles are being replicated, and the cell is bulking up on cytoplasm to make sure there's enough for both daughter cells. Mitosis may look like a waiting game, but there's plenty going on behind the scenes.

Waiting for action: Interphase

Interphase is the period when the cell isn't dividing. It begins when the new cells are done forming and ends when the cell prepares to divide. Although it's also called a "resting stage," there's constant activity in the cell during interphase.

Interphase is divided into subphases, each of which lasts anywhere from a few hours for those cells that divide frequently to days or years for those cells that divide less frequently (nerve cells, for example, can spend decades in interphase). The subphases are as follows:

- G_1, which stands for "gap" or "growth." During G_1, the cell creates its organelles, begins metabolism, grows, and synthesizes proteins.

- **S,** which stands for "synthesis." DNA synthesis or *replication* occurs during this subphase. The single double-helix DNA molecule inside the cell's nucleus becomes two new "sister" *chromatids,* and the centrosome is duplicated.

- G_2, which stands for "gap." Enzymes and proteins needed for cell division are produced during this subphase.

Sorting out the parts: Prophase

As the first active phase of mitosis, *prophase* is when structures in the cell's nucleus begin to disappear, including the nuclear membrane (or envelope), nucleoplasm, and nucleoli. The two centrioles that have formed from the centrosome push apart to opposite ends of the nucleus. Using protein filaments, they form poles and a *mitotic spindle* between them as well as *asters* (or *astral rays*) which radiate from the poles into the cytoplasm. At the same time, the *chromatin* threads (or *chromonemata*) shorten and coil, forming visible chromosomes. The chromosomes divide into chromatids that remain attached at an area called the *centromere,* which produces microtubules called *kinetochore* fibers. These interact with the spindle to assure that each daughter cell ultimately has a full set of chromosomes. The chromatids start to migrate toward the equatorial plane, an imaginary line between the poles.

Dividing at the equator: Metaphase

After the chromosomes are lined up and attached along the cell's newly formed equator, *metaphase* officially debuts. The nucleus itself is gone. The chromatids line up exactly along the center line of the cell (or the equatorial plane), attaching to the mitotic spindle by the centromere. The centromere also is attached by microtubules to opposite poles of the cell.

Packing up to move out: Anaphase

In *anaphase,* the centromeres split, separating the duplicate chromatids and forming two chromosomes. The spindles attached to the divided centromeres shorten, pulling the chromosomes toward the opposite poles. The cell begins to elongate. In late anaphase, as the chromosomes approach the poles, a slight furrow develops in the cytoplasm, showing where cytokinesis will eventually take place.

Pinching off: Telophase

Telophase occurs as the cell nears the end of division. The spindles and asters of early mitosis disappear, and each newly forming cell begins to synthesize its own structure. New nuclear membranes enclose the separated chromosomes. The coiled chromosomes unwind, becoming chromonemata once again. There's a more pronounced pinching, or furrowing, of the cytoplasm into two separate bodies, but there continues to be only one cell.

Splitting up: Cytokinesis

Cytokinesis means it's time for the big break-up. The furrow intended to divide the newly formed sister nuclei at last gets to finish the job. It migrates inward until it cleaves the single, altered cell into two new cells. Each new cell is smaller and contains less cytoplasm than the mother cell, but the daughter cells are genetically identical to each other and to the original mother cell.

Try this warm-up question on cell division:

Q. Cell division takes place to

    a. Repair injuries

    b. Replace nonfunctioning cells

    c. Grow the organism

    d. All of the above

A. The correct answer is all of the above. In addition, single-cell organisms use cell division for asexual reproduction.

1. Cells are dormant during interphase.

    a. True

    b. False

2. The G_1 subphase of interphase is

    a. The period of DNA synthesis

    b. The most active phase

    c. The phase between S and G_2

    d. Part of cell division

3. DNA is duplicated during which subphase?

    a. S

    b. G_2

    c. B

    d. G_1

4. The nuclear membrane, or envelope, disappears during

   a. Telophase

   b. Metaphase

   c. Prophase

   d. Interphase

5. Which of the following happens in prophase?

   a. The chromatids align on the equatorial plane.

   b. The chromosomes divide into chromatids.

   c. The nucleus reappears.

   d. The chromosomes move to opposite poles.

6. Which of the following is true for metaphase?

   a. The nuclear membrane appears.

   b. The chromosomes move to the poles.

   c. The chromatids align on the equatorial plane.

   d. It's composed of subphases G_1, S, and G_2.

7. During metaphase, each chromosome consists of two duplicate chromatids.

   a. True

   b. False

8. Identify an event that does *not* happen during anaphase.

   a. Early cytokinesis occurs with slight furrowing.

   b. The cell goes through subphase G_1.

   c. Spindles shorten.

   d. The centromeres split.

9. Genetically identical chromosomes are pulled toward opposite poles during

   a. Telophase

   b. Metaphase

   c. Anaphase

   d. Interphase

10. Which event does *not* occur during telophase?

   a. The chromosomes uncoil.

   b. The chromosomes reach the poles.

   c. The chromosomes become more distinct.

   d. The nuclear membrane reforms.

11. What structures disappear during telophase?

   a. Spindles and asters

   b. Nuclear membranes

   c. Nucleolei

   d. Chromonemata

12. Which is the correct order of mitosis?

    a. Prophase, interphase, metaphase, telophase, anaphase

    b. Interphase, prophase, metaphase, anaphase, telophase

    c. Metaphase, anaphase, telophase, interphase, prophase

    d. Anaphase, metaphase, telophase, interphase, prophase

13.–24. Use the terms that follow to identify the stages and cell structures shown in Figure 3-1.

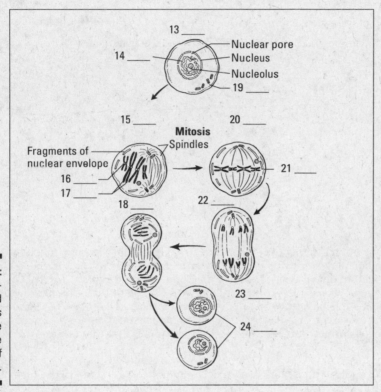

Figure 3-1: Cell structures and changes that make up the stages of mitosis.

    a. Anaphase

    b. Centromere

    c. Daughter cells

    d. Chromatin

    e. Cytokinesis

    f. Telophase

    g. Interphase

    h. Chromosomes aligned at equator

    i. Metaphase

    j. Centrioles

    k. Prophase

    l. Chromatids

25. The two newly formed daughter cells are

    a. The same size as the mother cell

    b. Not genetically identical to each other

    c. Unequal in size

    d. Genetically identical to the mother cell

26. Cytokinesis can be described as

    a. The period of preparation for cell division

    b. The dividing of the cytoplasm to surround the two newly formed nuclei

    c. The stage of alignment of the chromatids on the equatorial plane

    d. The initiation of cell division

27. Cytokinesis occurs during

    a. Telophase

    b. Interphase

    c. Prophase

    d. Metaphase

What Can Go Wrong

With the millions upon millions of cell divisions that happen in the human body, it's not surprising that sometimes things go wrong. An error during mitosis is called a *mutation.* One kind of mutation is *nondisjunction,* or a failure to separate. In this mutation, newly formed chromosomes don't quite divide, leaving one daughter cell with one more chromosome than normal and the other daughter cell one chromosome shy of a full complement. Down's syndrome is an example of what happens when nondisjunction occurs. A normal human cell has 46 chromosomes, but that of a Down's sufferer has 47.

Mitosis can also end up on fast-forward. Accelerated mitosis can lead to the formation of a tumor, also called a *neoplasm.* The rate of division usually restricts itself to replacing worn out or injured cells, but with accelerated mitosis, the cells don't know when to stop dividing.

28. Any change in a cell's genetic information is known as

    a. Phagocytosis

    b. Mutation

    c. Pinocytosis

    d. Neurotransmission

29.–33. Match the mitotic cell division stage with the appropriate activity.

29. _____ Chromatids line up along the equatorial plane

30. _____ Chromosomes contract and divide into chromatids

31. _____ Nondividing nucleus

32. _____ Chromosomes enclosed again in nuclear membrane at each pole

33. _____ Chromosomes attached to spindles, moving to opposite ends of a molecule

a. Anaphase
b. Prophase
c. Metaphase
d. Interphase
e. Telophase

34. Use the space provided to draw a basic illustration of each of the six stages inside a cell during mitosis.

Late interphase (1)	Prophase (2)
Metaphase (3)	Anaphase (4)
Telophase (5)	Cytokinesis (6)

Answers to Questions on Mitosis

The following are answers to the practice questions presented in this chapter.

1 Cells are dormant during interphase. **b. False.** Cells are at their most active during interphase.

2 The G_1 subphase of interphase is **b. the most active phase.** With all that organelle-growing, metabolizing, and protein-synthesizing that takes place during G_1, this isn't surprising.

3 DNA is duplicated during which subphase? **a. S.** Remember, the cell is S-ynthesizing new DNA molecules during this phase.

4 The nuclear membrane, or envelope, disappears during **c. prophase.**

5 Which of the following happens in prophase? **b. The chromosomes divide into chromatids.** But don't forget that they remain attached at the centromere.

6 Which of the following is true for metaphase? **c. The chromatids align on the equatorial plane.** Each of the other answer choices occurs during earlier or later phases.

7 During metaphase, each chromosome consists of two duplicate chromatids. **a. True.** That way, each resulting daughter cell will have identical chromosomes.

8 Identify an event that does *not* happen during anaphase. **b. The cell goes through subphase G_1.** G_1 took place back in interphase.

9 Genetically identical chromosomes are pulled toward opposite poles during **c. anaphase.** As the cell nears the end of division, it makes sense that duplicate packages move to opposite ends of the cell.

10 Which event does *not* occur during telophase? **c. The chromosomes become more distinct.** That change happened back in interphase.

11 What structures disappear during telophase? **a. Spindles and asters.** These structures disappear because they're no longer needed at the end of mitosis.

12 Which is the correct order of mitosis? **b. Interphase, prophase, metaphase, anaphase, telophase.**

13 – 24 Following is how Figure 3-1, the stages and structures of mitosis, should be labeled.

13. **g. Interphase**; 14. **d. Chromatin**; 15. **k. Prophase**; 16. **b. Centromere**; 17. **l. Chromatids**; 18. **f. Telophase**; 19. **j. Centrioles**; 20. **i. Metaphase**; 21. **h. Chromosomes aligned at equator**; 22. **a. Anaphase**; 23. **e. Cytokinesis**; 24. **c. Daughter cells**

25 The two newly formed daughter cells are **d. genetically identical to the mother cell.** None of the other answer options make sense.

26 Cytokinesis can be described as **b. the dividing of the cytoplasm to surround the two newly formed nuclei.**

27 Cytokinesis occurs during **a. telophase.** No, it's not a trick question. We told you early on that mitosis and cytokinesis ended at the same time. That means, of course, that cytokinesis takes place during the final stage of mitosis.

28 Any change in a cell's genetic information is known as **b. mutation.**

29 Chromatids line up along the equatorial plane: **c. Metaphase**

30 Chromosomes contract and divide into chromatids: **b. Prophase**

31 Nondividing nucleus: **d. Interphase**

32 Chromosomes enclosed again in nuclear membrane at each pole: **e. Telophase**

33 Chromosomes attached to spindles, moving to opposite ends of a molecule: **a. Anaphase**

34 Check your answers to the fill-in-the-blanks in questions 13 through 24 (Figure 3-1), and then compare your drawings to that figure.

Chapter 4

The Study of Tissues: Histology

*O*h, what tangled webs we weave! As the chapter title says, *histology* is the study of tissues, but you may be surprised to find out that the Greek *histo* doesn't translate as "tissue" but instead as "web." It's a logical next step after reviewing the cell and cellular division to take a look at what happens when groups of similar cells "web" together to form tissues. The four different types of tissue in the body are as follows:

✔ Epithelial, or skin, tissue (from the Greek *epi–* for "over" or "outer")

✔ Connective tissue

✔ Muscle tissue

✔ Nerve tissue

In this chapter, you find a quick review of the basics of each of these types of tissues along with practice questions to test your knowledge of them.

Getting Under Your Skin

Perhaps because of its unique job of both protecting the outer body and lining internal organs, epithelial tissue comes in more varieties than any other tissue.

Epithelial tissues, which generally are arranged in sheets or tubes of tightly-packed cells, always have a free, or *apical,* surface that can be exposed to the air or to fluid. That free surface also can be covered by additional layers of epithelial tissue. But whether it's layered or not, each epithelial cell has *polarity* (a top and a bottom), and all but one side of the cell is tucked snugly against neighboring cells. The apical side sometimes has cytoplasmic projections such as *cilia,* hair-like growths that can move material over the cell's surface, or *microvilli,* finger-like projections that increase the cell's surface area for absorption. Opposite the apical side is the *basal* side (think basement), which typically attaches to some kind of connective tissue.

Epithelial tissue serves several key functions, including the following:

✔ **Protection:** Skin protects vulnerable structures or tissues deeper in the body.

✔ **Barrier:** Epithelial tissues prevent foreign materials from getting inside the body.

✔ **Sensation:** Sensory nerve endings embedded in epithelial tissue connect the body with outside stimuli.

✔ **Secretion:** Epithelial tissue in glands can be specialized to secrete enzymes, hormones, and fluids.

Single-layer epithelial tissue is classified as *simple*. Tissue with more than one layer is called *stratified*. Epithelial tissues also can be classified according to shape: *Squamous* is a thin, flat cell; *cuboidal* is, as the name implies, equal in height and width and shaped like a cube; and *columnar* cells are taller than they are wide.

Following are the ten primary types of epithelial tissues:

✔ **Simple squamous epithelium:** Looking a bit like rolling tundra, this flat layer of scale-like cells is useful in diffusion, secretion, or absorption. Each cell nucleus is centrally located and is round or oval. Simple squamous epithelium lines the lungs' air sacs where oxygen and carbon dioxide are exchanged; forms blood filters inside the kidneys; and lines the inner surface of the eardrum, known as the tympanic membrane.

✔ **Simple cuboidal epithelium:** These cube-shaped cells, found in a single layer that looks like a microscopic mattress, have centrally located nuclei that usually are round. Found in the ovaries, kidneys, and some glands, this type of epithelium functions in secretion, absorption, and tube formation.

✔ **Simple columnar epithelium:** These densely packed cells are taller than they are wide, with nuclei located near the base of each cell. Found lining the digestive tract from the stomach to the anal canal, this type of epithelium functions in secretion and absorption.

✔ **Simple columnar ciliated epithelium:** A close cousin to simple columnar epithelium, this type of tissue has hair-like cilia that can move mucus and other substances across the cell. It's found lining the small respiratory tubes.

✔ **Pseudostratified columnar epithelium:** Pay attention to the prefix *pseudo*– here, which means "false." It may look multilayered because the cells' nuclei are scattered at different levels, but it's not. This type of epithelium is found in the salivary glands and some segments of the male reproductive system, including the urethra.

✔ **Pseudostratified columnar ciliated epithelium:** Another variation on a theme, this tissue is nearly identical to pseudostratified columnar epithelium. The difference is that this tissue's free surface has cilia, making it ideal for lining air passages because the cilia's uniform waving action causes a thin layer of mucus to move in one direction — toward the throat and mouth — and trap dust particles.

✔ **Stratified squamous epithelium:** This tissue is the stuff you see everyday — your outer skin, or epidermis. This multilayered tissue has squamous cells on the outside plus deeper layers of cuboidal or columnar cells. Found in areas where the outer cell layer is constantly worn away, this type of epithelium regenerates its surface layer with cells from lower layers.

✔ **Stratified cuboidal epithelium:** This multilayered epithelium can be found in sweat glands, conjunctiva of the eye, and the male urethra. Its function is primarily protection.

✔ **Stratified columnar epithelium:** Also multilayered, this epithelium is found lining parts of the male urethra, excretory ducts of glands, and some small areas of the anal mucus membrane.

✔ **Stratified transitional epithelium:** This epithelium is referred to as *transitional* because its cells can shape-shift from cubes to squamous-like flat surfaces and back again. Found lining the bladder, the cells flatten out to make room for urine.

Following are some practice questions dealing with epithelial tissue:

EXAMPLE

0. Stratified epithelial tissue can be described as

a. A thin sheet of cells

b. Covered in cilia

c. Layers of stacked epithelial cells

d. A long string of tissue

A. The correct answer is layers of stacked epithelial cells. Remember that "stratified" means layers.

1. Epithelial cells can be shaped

a. Like columns

b. Like cubes

c. Thin and flat

d. All of the above

2. Epithelial tissue is classified by

a. Number of layers

b. Composition of matrix

c. Cell shape

d. Both the number of layers and the cell shape

3. The epithelial tissue that has the ability to stretch is

a. Simple squamous

b. Transitional

c. Pseudostratified columnar

d. Simple columnar

4.–8. Match the epithelial tissue with its location in the body.

4. _____ Simple columnar

5. _____ Stratified squamous

6. _____ Transitional

7. _____ Pseudostratified columnar ciliated

8. _____ Simple cuboidal

a. Urinary bladder

b. Tubules of the kidney

c. Digestive tract

d. Epidermis of the skin

e. Respiratory passages

9. A tissue that's one layer thick but appears to be multilayered and is composed of cells taller than they are wide is

a. Stratified ciliated columnar epithelium

b. Simple squamous epithelium

c. Pseudostratified columnar epithelium

d. Transitional epithelium

10.–19. Use the terms that follow to identify the epithelial tissues shown in Figure 4-1.

CELL SHAPES   SIMPLE   STRATIFIED

10. _____

11. _____

12. _____

13. _____

14. _____

15. _____

16. _____

17. _____

18. _____

19. _____

Figure 4-1:
Epithelial
tissues.

Illustration by Imagineering Media Services Inc.

a. Stratified squamous

b. Simple columnar

c. Squamous

d. Transitional stretched

e. Simple squamous

f. Columnar

g. Pseudostratified

h. Cuboidal

i. Transitional relaxed

j. Simple cuboidal

Making a Connection: Connective Tissue

Connective tissues connect, support, and bind body structures together. Unlike other types of tissues, connective tissues are classified more by the stuff in which the cells lay — the extracellular matrix — than by the cells themselves. The cells that produce that matrix are scattered within it like chocolate chips in ice cream. The load-bearing strength of connective tissue comes from a fibrous protein called *collagen.* All connective tissues contain a varying mix of collagen, elastic, and reticular fibers.

Following are the primary types of connective tissue:

- **Areolar, or loose, tissue:** This tissue exists between and around almost everything in the body to bind structures together and fill space. It's made up of wavy ribbons called *collagenous protein fibers,* cylindrical threads called *elastic fibers,* and *amorphous ground substance,* a semisolid gel. Various cells including lymphocytes, fibroblasts, fat cells, and mast cells are scattered throughout the ground substance (see Figure 4-2).

- **Dense regular connective tissue:** Made up of parallel, densely packed bands or sheets of fibers (see Figure 4-2), this type of tissue is found in tendons as bundles of collagenous fibers attaching muscles to bone and in ligaments as bundles of elastic fibers extending from bone to bone, surrounding a joint, and anchoring organs. It usually resists force in just two directions.

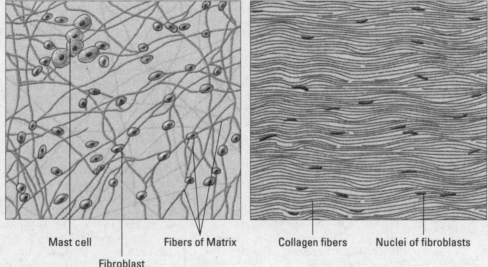

Figure 4-2: Areolar tissue and dense regular connective tissue.

Mast cell   Fibers of Matrix   Collagen fibers   Nuclei of fibroblasts

Fibroblast

Illustration by Imagineering Media Services Inc.

- **Dense irregular connective tissue:** Also known as *dense fibrous connective tissue,* it consists of fibers that twist and weave around each other, forming a thick tissue that can withstand stresses applied from any direction. This tissue makes up the strong inner skin layer called the *dermis* as well as the outer capsule of organs like the kidney and the spleen.

- **Adipose tissue:** Composed of fat cells, this tissue forms padding around internal organs, reduces heat loss through the skin, and stores energy in fat molecules called triglycerides. Fat molecules fill the cells, forcing the nuclei against the cell membranes and giving them a ring-like shape. Adipose has an *intracellular matrix* rather than an extracellular matrix.

- **Reticular tissue:** Literally translated as "web-like" or "net-like," reticular tissue is made up of slender, branching reticular fibers with reticular cells overlaying them. Its intricate structure makes it a particularly good filter, which explains why it's found inside the spleen, lymph nodes, and bone marrow.

✔ **Cartilage:** These firm but flexible tissues, made up of collagen and elastic fibers, have no blood vessels or nerve cells (a state called *non-vascular* or *avascular*). Cartilage contains openings called *lacunae* (from the Latin word *lacus* for "lake" or "pit") that enclose mature cells called *chondrocytes,* which are preceded by cells called *chondroblasts.* A membrane known as the *perichondrium* surrounds cartilage tissue, which also contains a gelatinous protein called *chondrin.* There are three types of cartilage:

- **Hyaline cartilage:** The most abundant cartilage in the body, it's elastic and made up of a uniform matrix pocked with chondrocytes. It lays the foundation for the embryonic skeleton, forms the rib (or *costal*) cartilages, makes up nose cartilage, and covers the articulating surfaces of bones.

- **Fibrocartilage:** As the name implies, fibrocartilage contains thick, compact collagen fibers. The sponge-like structure, with the lacunae and chondrocytes lined up within the fibers, makes it a good shock absorber. It's found in the intervertebral discs of the vertebral column and in the symphysis pubis at the front of the pelvis.

- **Elastic cartilage:** Similar to hyaline cartilage, elastic cartilage has more tightly packed lacunae and chondrocytes between parallel elastic fibers. This structure, which makes up the ear lobe and other structures where a specific form is important, tends to bounce back to its original shape after being bent.

✔ **Bone, or osseous, tissue:** Essentially, bone is mineralized connective tissue formed into repeating patterns called *Haversian systems.* In the center of each system is a large opening, the *Haversian canal,* that contains blood vessels, lymph vessels, and nerves. The central canal is surrounded by thin membranes called *lamellae* that contain the lacunae, which in turn contain *osteocytes* (bone cells). Smaller *canaliculi* connect the lacunae and circulate tissue fluids from the blood vessels to nourish the osteocytes. (We explore bone in more detail in Chapter 5.)

✔ **Blood:** Yes, blood is considered a type of connective tissue. Like other connective tissues, it has an extracellular matrix — in this case, plasma — in which are suspended *erythrocytes* (red blood cells), *leukocytes* (white blood cells), and *thrombocytes* (platelets). (Blood also is considered a vascular tissue because it circulates inside arteries and veins, but we get into that in Chapter 10.) Roughly half of blood's volume is fluid or plasma while the other half is suspended cells. Erythrocytes are concave on both sides and contain a pigment, *hemoglobin,* which supplies oxygen to the body's cells and takes carbon dioxide away. There are approximately 5 million erythrocytes per cubic millimeter of whole blood. Thrombocytes, which number approximately 250,000 per cubic millimeter, are fragments of cells used in blood clotting. Leukocytes are large *phagocytic* cells (literally "cell that eats") that are part of the body's immune system. There are, however, relatively few of them — less than 10,000 per cubic millimeter.

20. Adipose tissue is composed of

   a. Mast cells

   b. Chondrocytes

   c. Osteocytes

   d. Fat cells

21. Tendons are composed of

   a. Elastic tissue

   b. Dense regular connective tissue

   c. Areolar connective tissue

   d. Fibrocartilage

22. The tissue covering the surface of articulating bones is

  a. Hyaline cartilage

  b. Areolar

  c. Vascular tissue

  d. Fibrocartilage

23. Vascular connective tissue is

  a. Hyaline cartilage

  b. Elastic tissue

  c. Blood

  d. Bone

24. Tissue containing lacunae with osteocytes is

  a. Elastic cartilage

  b. Bone

  c. Hyaline cartilage

  d. Blood

25. Blood contains cells functional in clotting called

  a. Phagocytes

  b. Erythrocytes

  c. Leukocytes

  d. Thrombocytes

Flexing It: Muscle Tissue

Although we review how muscles work in Chapter 6, in histology you should know that muscle tissue is made up of fibers known as *myocytes*. The cytoplasm within the fibers is called *sarcoplasm,* and within that sarcoplasm are minute *myofibrils* that contain the protein filaments *actin* and *myosin*. These filaments slide past each other during a muscle contraction, shortening the fiber.

Following are the three types of muscle tissue (see Figure 4-3):

  ✔ **Smooth muscle tissue:** This type of tissue contracts without conscious control. Made up of spindle-shaped fibers with large, centrally located nuclei, it's found in the walls of internal organs, or *viscera*. Smooth muscle gets its name from the fact that, unlike other muscle tissue types, it is not striated.

  ✔ **Cardiac muscle tissue:** Also known as *myocardium,* cardiac muscle tissue is made of branching fibers, each with a central nucleus and alternating light and dark striations. Between the fibers are dark structures called *intercalated discs*. As with smooth muscle, cardiac muscle tissue contractions occur through the autonomic nervous system (involuntary control).

  ✔ **Skeletal, or striated, muscle tissue:** Biceps, triceps, pecs — these are the muscles that bodybuilders focus on. As the name implies, skeletal muscles attach to the skeleton and are used throughout the central nervous system for movement. Muscle fibers are cylindrical with several nuclei in each cell (which makes them *multinucleated*) and cross-striations throughout.

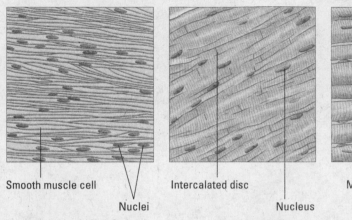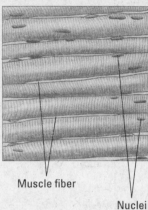

Figure 4-3:
Muscle
tissues:
Smooth,
cardiac, and
skeletal.

Smooth muscle cell   Intercalated disc   Muscle fiber

Nuclei   Nucleus   Nuclei

Illustration by Imagineering Media Services Inc.

26. Which type of tissue is multinucleated?

   a. Skeletal muscle tissue

   b. Cardiac muscle tissue

   c. Smooth muscle tissue

27. A tissue that has intercalated discs is

   a. Cardiac muscle

   b. Skeletal muscle

   c. Smooth muscle

   d. Striated muscle

28. Skeletal muscle tissue has prominent lines across the fiber called

   a. Fibroblasts

   b. Multinucleation

   c. Lacunae

   d. Striations

29. Smooth muscle tissue is found in the

   a. Heart

   b. Urinary bladder

   c. Bicep

   d. Deltoid

Getting the Signal Across: Nerve Tissue

There's only one type of nerve tissue and only one primary type of cell in it: the
neuron. Nerve tissue is unique in that it can both generate and conduct electrical signals in the body. That process starts when sense receptors receive a stimulus that
causes electrical impulses to be sent through finger-like cytoplasmic projections called

dendrites. From there, the impulse moves through the body of the cell and into another type of cytoplasmic projection (or nerve process) called an *axon* that hands the signal off to the next cell down the line. (We look more closely at how all that happens when we examine the central nervous system in Chapter 15.)

Following are some practice questions dealing with nerve tissue:

30. Cells capable of producing and transmitting electrical impulses are

   a. Schwann cells

   b. Neurons

   c. Chondrocytes

   d. Thrombocytes

31. The cytoplasmic projection of a neuron that carries impulses away from the cell body is called

   a. A myofibril

   b. A dendrite

   c. An axon

   d. A cross-striation

32. The cytoplasmic projections that receive stimuli from sense receptors are

   a. Dendrites

   b. Collagenous fibers

   c. Axons

   d. Schwann projections

33.–37. Match each description with the appropriate tissue.

33. _____ Precedes bone formation in embryonic development   **a.** Areolar tissue

34. _____ Found in visceral walls   **b.** Hyaline cartilage

35. _____ Found in and around most structures in the body   **c.** Bone

36. _____ Found in the external ear   **d.** Elastic cartilage

37. _____ Supports soft tissues of the body   **e.** Smooth muscle

Answers to Questions on Histology

The following are answers to the practice questions presented in this chapter.

1. Epithelial cells can be shaped **d. all of the above.** Cube, column, or flat and thin, it's still an epithelial cell.

2. Epithelial tissue is classified by **d. both the number of layers and the cell shape.** Classification requires looking at both simultaneously.

3. The epithelial tissue that has the ability to stretch is **b. transitional.** This tissue lines the bladder, so it had better be stretchy!

4. Simple columnar: **c. Digestive tract**

5. Stratified squamous: **d. Epidermis of the skin**

6. Transitional: **a. Urinary bladder**

7. Pseudostratified columnar ciliated: **e. Respiratory passages**

8. Simple cuboidal: **b. Tubules of the kidney**

9. A tissue that's one layer thick but appears to be multilayered and is composed of cells taller than they are wide is **c. pseudostratified columnar epithelium.** To arrive at the correct answer, consider this question one piece at a time: *pseudo* is "false," *stratified* means "layered" (so you have "false-layered"), and columns are taller than they are wide.

10-19. Following is how Figure 4-1, the types of epithelial cells and tissues, should be labeled.

> 10. **c. Squamous**; 11. **h. Cuboidal**; 12. **f. Columnar**; 13. **e. Simple squamous**; 14. **j. Simple cuboidal**; 15. **b. Simple columnar**; 16. **g. Pseudostratified**; 17. **a. Stratified squamous**; 18. **i. Transitional relaxed**; 19. **d. Transitional stretched**

20. Adipose tissue is composed of **d. fat cells.** Think of the Latin *adeps,* which means "fat."

21. Tendons are composed of **b. dense regular connective tissue.** Tendons are dense and exhibit a regular pattern.

22. The tissue covering the surface of articulating bones is **a. hyaline cartilage.**

23. Vascular connective tissue is **c. blood.**

24. Tissue containing lacunae with osteocytes is **b. bone.** Remember that *osteo* is the Latin word for "bone."

25. Blood contains cells functional in clotting called **d. thrombocytes.** Knowing that the Greek word *thrombos* means "clot" can help you spot the correct answer in this question.

26. Which type of tissue is multinucleated? **a. Skeletal muscle tissue.** The other answer options are tissues that have only one nucleus per cell.

27. A tissue that has intercalated discs is **a. cardiac muscle.** Intercalated discs, as you should or will know from studying the circulatory system, are involved in conducting signals for the heart to pump.

28 Skeletal muscle tissue has prominent lines across the fiber called **d. striations.** Striations are light and dark lines across the fiber.

29 Smooth muscle tissue is found in the **b. urinary bladder.** The other answer choices contain striated tissue, which technically means that they aren't smooth.

30 Cells capable of producing and transmitting electrical impulses are **b. neurons.** These are the cells that make up nerve tissue.

31 The cytoplasmic projection of a neuron that carries impulses away from the cell body is called **c. an axon.** Each neuron cell usually has only one axon, although it may branch off several times.

32 The cytoplasmic projections that receive stimuli from sense receptors are **a. dendrites.** Usually, there are several dendrites per neuron cell.

33 Precedes bone formation in embryonic development: **b. Hyaline cartilage**

34 Found in visceral walls: **e. Smooth muscle**

35 Found in and around most structures in the body: **a. Areolar tissue**

36 Found in the external ear: **d. Elastic cartilage**

37 Supports soft tissues of the body: **c. Bone**

Part II
Weaving It Together: Bones, Muscles, and Skin

The 5th Wave   By Rich Tennant

"Your bones are in terrific shape. However, your husband's wood rot will require a whole new stake being shoved down his back."

In this part . . .

This part gets into what most people think of first when it comes to anatomy and physiology: bones and muscles. First we focus on how bones are formed before broadening the view to the axial skeleton (the parts that line up from head to toe) and the appendicular skeleton (the parts that reach out from the central axis). You review how muscles attach to that framework and watch the body take shape before wrapping this newly layered package in the body's largest single organ: the skin.

Chapter 5

A Scaffold to Build On: The Skeleton

Human *osteology,* from the Greek word for "bone" *(osteon)* and the suffix *–logy,* which means "to study," focuses on the 206 bones in the adult body endoskeleton. But it's more than just bones; it's also ligaments and cartilage and the joints that make the whole assembly useful. In this chapter, you get lots of practice exploring the skeletal functions and how the joints work together.

Understanding Dem Bones

The skeletal system as a whole serves five key functions:

- **Protection:** The skeleton encases and shields delicate internal organs that might otherwise be damaged during motion or crushed by the weight of the body itself. For example, the skull's cranium houses the brain, and the ribs and sternum of the thoracic cage protect organs in the central body cavity.

- **Movement:** By providing anchor sites and a scaffold against which muscles can contract, the skeleton makes motion possible. The bones act as levers, the joints are the fulcrums, and the muscles apply the force. For instance, when the biceps muscle contracts, the radius and ulna bones of the forearm are lifted toward the humerus bone of the upper arm.

- **Support:** The vertebral column's curvatures play a key role in supporting the entire body's weight, as do the arches formed by the bones of the feet. Upper body support flows from the *clavicle,* or collarbone, which is the only bone that attaches the upper extremities to the axial skeleton and the only horizontal long bone in the human body.

- **Mineral storage:** Calcium, phosphorous, and other minerals like magnesium must be maintained in the bloodstream at a constant level, so they're "banked" in the bones in case the dietary intake of those minerals drops. The bones' mineral content is constantly renewed, refreshing entirely about every nine months. A 35 percent decrease in blood calcium will cause convulsions.

- **Blood cell formation:** Called *hemopoiesis* or *hematopoiesis,* most blood cell formation takes place within the red marrow inside the ends of long bones as well as within the vertebrae, ribs, sternum, and cranial bones. Marrow produces three types of blood cells: *erythrocytes* (red cells), *leukocytes* (white cells), and *thrombocytes* (platelets). Most of these are formed in red bone marrow, although some types of white blood cells are produced in fat-rich yellow bone marrow. At birth, all bone marrow is red. With age, it converts to the yellow type. In cases of severe blood loss, the body can convert yellow marrow back to red marrow in order to increase blood cell production.

The following are examples of questions dealing with skeletal functions:

Q. Which of the following is not a function of the skeleton?

a. Support of soft tissue

b. Hemostasis

c. Production of red blood cells

d. Movement

A. The correct answer, of course, is hemostasis, which is the stoppage of bleeding or blood flow. (We discuss hemostasis further when we address the circulatory system in Chapter 10.) This is one of those frequent times when study of anatomy and physiology boils down to rote memorization of Latin and Greek roots (check out the Cheat Sheet at the front of this book).

Q. The skeletal system is composed of

a. Bones

b. Cartilage

c. Joints

d. All of the above

A. If you hesitated to choose "all of the above," ask yourself this: If you suspected one of the three things was *not* part of the skeletal system, to which system would it belong? Can't think of a sensible alternative? Neither can we.

1. The formation of red blood cells by the bone marrow is known as

a. Hemolysis

b. Hematopoiesis

c. Hemoptysis

d. Hematuria

2. The following mineral is stored inside bones for later use:

a. Phosphorous

b. Calcium

c. Magnesium

d. All of the above

3. Besides support and protection, the skeleton serves other important functions, including

a. Reproduction

b. Locomotion

c. Respiration

d. Circulation

4. The curvatures in some bone structures serve the following purpose:

a. Support the body

b. Make the body more flexible

c. Enhance circulation throughout the body

d. Divide different body areas

Boning Up on Classifications, Structures, and Ossification

Adult bones are composed of 30 percent protein (called *ossein*), 45 percent minerals (including calcium, phosphorus, and magnesium), and 25 percent water. Minerals give the bone strength and hardness. At birth, bones are soft and pliable because of cartilage in their structure. As the body grows, older cartilage gradually is replaced by hard bone tissue. Mineral in the bones increases with age, causing them to become more brittle and easily fractured.

Various types of bone make up the human skeleton, but fortunately for memorization purposes, bone type names match what the bones look like. They are as follows:

- **Long bones,** like those found in the arms and legs, form the weight-bearing part of the skeleton.

- **Short bones,** such as those in the wrists (carpals) and ankles (tarsals), have a blocky structure and allow for a greater range of motion.

- **Flat bones,** such as the skull, sternum, scapulae, and pelvic bones, shield soft tissues.

- **Irregular bones,** such as the mandible (jawbone) and vertebrae, come in a variety of shapes and sizes suited for attachment to muscles, tendons, and ligaments. Irregular bones include seed-shaped sesamoid bones found in joints such as the patella, or kneecap.

Unfortunately for students of bone structures, there's no easy way to memorize them. So brace yourself for a rapid summary of what your textbook probably goes into in much greater detail.

Compact bone is a dense layer made up of structural units, or *lacunae,* arranged in concentric circles called *Haversian systems* (also referred to in short as *osteons*), each of which has a central, microscopic *Haversian canal.* A perpendicular system of canals, called *Volkmann's canals,* penetrate and cross between the Haversian systems. This network ensures circulation into even the hardest bone structure. Compact bone tissue is thick in the shaft and tapers to paper thinness at the ends of the bones. The bulbous ends of each long bone, known as the *epiphyses* (or singularly as an *epiphysis*), are made up of spongy bone or cancellous bone tissue covered by a thin layer of compact bone. The *diaphysis,* or shaft, contains the *medullary cavity* and blood cell–producing marrow. A membrane called the *periosteum* covers the outer bone to provide nutrients and oxygen, remove waste, and connect with ligaments and tendons.

Bones grow through the cellular activities of *osteoblasts* on the surface of the bone, which produce layers of mature bone cells called *osteocytes. Osteoclasts* are cells that function in the developing fetus to absorb cartilage as ossification occurs and function in adult bone to break down and remove spent bone tissue.

There are two types of *ossification,* which is the process by which softer tissues harden into bone. Both types rely on a peptide hormone produced by the thyroid gland, *calcitonin,* which regulates metabolism of calcium, the body's most abundant mineral. The two types of ossification are

- **Endochondral or intracartilaginous ossification:** Occurs when mineral salts, particularly calcium and phosphorus, calcify along the scaffolding of cartilage formed in the developing fetus beginning about the fifth week after conception. This process, known as *calcification,* takes place in the presence of vitamin D and a hormone from the parathyroid gland. The absence of any one of these substances causes a child to have soft bone, called *rickets.* Next, the blood supply entering the cartilage brings osteoblasts that attach themselves to the cartilage. As the primary center of ossification, the diaphysis of the long bone is the first to form spongy bone tissue along the cartilage, followed by the epiphyses, which form the secondary centers of ossification and are separated from the diaphysis by a layer of uncalcified cartilage called the *epiphyseal plate* where all growth in bone length occurs. Compact bone tissue covering the bone's surface is produced by osteoblasts in the inner layer of the periosteum, producing growth in diameter.

✔ **Intramembranous ossification:** Occurs not along cartilage but instead along a template of membrane, as the name implies, primarily in compact flat bones of the skull that don't have Haversian systems. The skull and mandible (lower jaw) of the fetus are first laid down as a membrane. Osteoblasts entering with the blood supply attach to the membrane, ossifying from the center of the bone outward. The edges of the skull's bones don't completely ossify to allow for molding of the head during birth. Instead, six soft spots, or *fontanels,* are formed: one frontal or anterior, two sphenoidal or anterolateral, two mastoidal or posterolateral, and one occipital or posterior.

Once formed, bone is surrounded by the periosteum, which has both a vascular layer (remember the Latin word for "vessel" is *vasculum*) and an inner layer that contains the osteoblasts needed for bone growth and repair. A penetrating matrix of connective tissue called *Sharpey's fibers* connects the periosteum to the bone; inside the bone, the medullary cavity is lined by a thin membrane called the *endosteum* (from the Greek *endon,* meaning "within," and, of course, that ever-present Greek word *osteon*).

Following are the basic terms used to identify bone landmarks or surface features:

✔ **Process:** A broad designation for any prominence or prolongation

✔ **Spine:** An abrupt or pointed projection

✔ **Trochanter:** A large, usually blunt process

✔ **Tubercle:** A smaller, rounded eminence

✔ **Tuberosity:** A large, often rough eminence

✔ **Crest:** A prominent ridge

✔ **Head:** A large, rounded articular end of a bone; often set off from the shaft by a neck

✔ **Condyle:** An oval articular prominence of a bone

✔ **Facet:** A smooth, flat or nearly flat articulating surface

✔ **Fossa:** A deeper depression

✔ **Sulcus:** A groove

✔ **Foramen:** A hole

✔ **Meatus:** A canal or opening to a canal

Q. Mature bone cells are called

a. Osteocytes

b. Osteoblasts

c. Chondrocytes

d. Osteoclasts

A. The correct answer is osteocytes. Why? Refer to the Latin root *cyta,* which is most commonly used to refer to a cell. Why not osteoblasts or osteoclasts? Because the Greek root *blast* in biological terms refers to growth or formation, and the Latin root *clast* refers to breaking or fragmentation.

Q. The basic unit of structure in adult compact bones is the

a. Osteoblast

b. Osteon

c. Osteoclast

d. Osteocytes

A. The correct answer is osteon. Always pay attention to exactly what a multiple-choice question is asking. Remember that description of the structural part of the bone, the Haversian system? And check out that root *osteo,* which comes from the Greek word for "bone."

5. The most abundant mineral in the body is

a. Potassium

b. Magnesium

c. Calcium

d. Sodium

6. A decrease in calcium concentration in the body fluids by 35 percent causes

a. Unresponsive neurons

b. Death

c. Convulsions

d. Muscle spasms

7. Calcitonin is produced by the

a. Thyroid gland

b. Pituitary gland

c. Adrenal gland

d. Parathyroid

8. Bones are encased by a membrane called the

a. Vasculum

b. Periosteum

c. Endosteum

d. Fontanel

9. A bone's encapsulating membrane is attached by

    a. Haversian canals

    b. Sphenoids

    c. Sharpey's fibers

    d. Mastoids

10. Bone marrow can be found inside the

    a. Medullary cavity

    b. Sharpey's cavity

    c. Ossifying cavity

    d. Diaphanous cavity

11. The six structures in the skull of an infant are called

    a. Condyls

    b. Fontanels

    c. Facets

    d. Trochanters

12. Volkmann's canals

    a. Are found in cancellous bone only

    b. Contain the nutrient artery

    c. Pass through the epiphysis

    d. Supply articulating cartilage blood

13. Blood vessels entering through Volkmann's canals reach the bone cells through the

    a. Endosteum

    b. Haversian canal

    c. Lacunae

    d. Medullary canal

14. The hormone released when calcium ion concentration is abnormally high is

    a. Thyroxine

    b. Calcitonin

    c. Progesterone

    d. Parathormone

15.–25. Fill in the blanks to complete the following sentences:

Bones are first laid down as **15.** _____ during the fifth week after conception. Development of the bone begins with **16.** _____, the depositing of calcium and phosphorus. Next, the blood supply entering the cartilage brings **17.** _____ that attach themselves to the cartilage. Ossification in long bones begins in the **18.** _____ of the long bone and moves toward the **19.** _____ of the bone. The epiphyseal and diaphyseal areas remain separated by a layer of uncalcified cartilage called the **20.** _____.

Another very large cell that enters with the blood supply is the **21.** _____, which helps absorb the cartilage as ossification occurs. Later it helps absorb bone tissue from the center of the long bone's shaft, forming the **22.** _____ cavity. After ossification, the spaces that were formed by the osteoclasts join together to form **23.** _____ systems, which contain the blood vessels, lymphatic vessels, and nerves. Unlike bones in the rest of the body, those of the skull and mandible (lower jaw) are first laid down as **24.** _____. In the skull, the edges of the bone don't ossify in the fetus but remain membranous and form **25.** _____.

26.–34. Classify the following bones by shape. Each classification may be used more than once.

26. _____ Vertebrae of the vertebral column   **a.** Flat bone

27. _____ Femur in thigh   **b.** Irregular bone

28. _____ Sternum   **c.** Long bone

29. _____ Tarsals in ankle   **d.** Short bone

30. _____ Humerus in upper arm

31. _____ Phalanges in fingers and toes

32. _____ Scapulae of shoulder

33. _____ Kneecap

34. _____ Carpals in wrist

35.–47. Match the description with the bone landmarks or surface features.

35. _____ An abrupt or pointed projection   **a.** Condyle

36. _____ A large, usually blunt process   **b.** Crest

37. _____ A designation for any prominence   **c.** Facet
          or prolongation
                                                      d. Foramen
38. _____ A large, often rough eminence
                                                      e. Fossa
39. _____ A prominent ridge
                                                      f. Head
40. _____ A large, rounded articular end of a bone; often
          set off from the shaft by the neck   **g.** Meatus

41. _____ An oval articular prominence of a bone   **h.** Process

42. _____ A smooth, flat or nearly flat articulating surface   **i.** Spine

43. _____ A deeper depression   **j.** Sulcus

44. _____ A groove   **k.** Trochanter

45. _____ A hole   **l.** Tubercle

46. _____ A canal or opening to a canal   **m.** Tuberosity

47. _____ A smaller, rounded eminence

48.–56. Use the terms that follow to identify the regions and structures of the long bone shown in Figure 5-1.

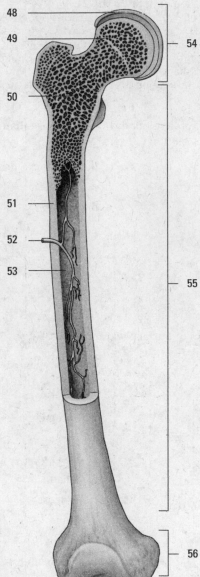

Figure 5-1:
The long
bone.

Illustration by Imagineering Media Services Inc.

a. Diaphysis

b. Medullary cavity

c. Distal epiphysis

d. Spongy bone tissue

e. Medullary or nutrient artery

f. Proximal epiphysis

g. Red bone marrow

h. Articulating cartilage

i. Compact bone tissue

Axial Skeleton: Keeping It All in Line

Just as the Earth rotates around its axis, the axial skeleton lies along the midline, or center, of the body. Think of your spinal column and the bones that connect directly to it — the rib (thoracic) cage and the skull. The tiny *hyoid* bone, which lies just above your larynx, or voice box, also is considered part of the axial skeleton, although it's the only bone in the entire body that doesn't connect, or articulate, with any other bone. It's also known as the *tongue bone* because the tongue's muscles attach to it.

There are a total of 80 named bones in the axial skeleton, which supports the head and trunk of the body and serves as an anchor for the pelvic girdle. Twenty-nine of those 80 bones are in (or very near) the skull. In addition to the hyoid bone, 8 bones form the cranium to house and protect the brain, 14 form the face, and 6 bones make it possible for you to hear.

Making a hard head harder

Fortunately for the cramming student, most of the bones in the skull come in pairs. In the cranium there's just one of each of the following: *frontal* bone (forehead), *occipital* bone (back and base of the skull), *ethmoid* bone (made of several plates, or sections, between the eye orbits in the nasal cavity), and *sphenoid* bone (a butterfly-shaped structure that forms the floor of the cranial cavity). But there are two *temporal* (housing the hearing organs in the *auditory meatus*) and *parietal* (roof and sides of the skull) bones. These bones are attached along *sutures* called *coronal* (located at the top of the skull), *squamosal* (located on the sides of the head surrounding the temporal bone), *sagittal* (along the midline atop the skull located between the two parietal bones), and *lambdoidal* (forming an upside-down V — the shape of the Greek letter lambda — on the back of the skull).

In the face, there's only one *mandible* (jawbone) and one *vomer* dividing the nostrils, but there are two each of *maxillary* (upper jaw), *zygomatic* (cheekbone), *nasal, lacrimal* (a small bone in the eye socket), *palatine* (which makes up part of the eye socket, nasal cavity, and roof of the mouth), and *inferior nasal concha,* or turbinated, bones. Inside the ear, there are two each of three ossicles, or bonelets, which also happen to be the smallest bones in the human body: the *malleus, incus,* and *stapes.*

The cranial cavity contains several openings, or *foramina* (the singular is *foramen*), in the floor of the cranial cavity that allow various nerves and vessels to connect to the brain. The holes in the ethmoid bone's *cribriform plate* allow olfactory — or sense of smell — receptors to pass through to the brain. A large hole in the occipital bone called the *foramen magnum* allows the spinal cord to connect with the brain. The sphenoid bone is riddled with foramina. The *optic foramen* allows passage of the optic nerves, whereas the *jugular foramen* allows passage of the jugular vein and several cranial nerves. The *foramen rotundum* allows passage of the *trigeminal nerve,* which is the chief sensory nerve to the face and controls the motor functions of chewing. The *foramen ovale* allows passage of the nerves controlling the tongue, among other things. The *foramen spinosum* allows passage of the middle *meningeal* artery, which supplies blood to various parts of the brain. The sphenoid bone also features the *sella turcica,* or Turk's saddle, that cradles the pituitary gland and forms part of the *foramen lacerum,* through which pass several key components of the autonomic nervous system.

Encased within the frontal, sphenoid, ethmoid, and maxillary bones of the skull are several air-filled, mucous-lined cavities called *paranasal sinuses*. While you may think their primary function is to drive you crazy with pressure and infections, the sinuses actually lighten the skull's weight, making it easier to hold your head up high; warm and humidify inhaled air; and act as resonance chambers to prolong and intensify the reverberations of your voice. *Mastoid sinuses* drain into the middle ear (hence the earache referred to as *mastoiditis*). *Maxillary sinuses* are flanked by the bones of the maxilla, the upper jaw. The *paranasal sinuses* are the ones that drain into the nose and cause so much trouble when you cry or have a cold.

Putting your backbones into it

The axial skeleton also consists of 33 bones in the vertebral column, laid out in four distinct curvatures, or areas.

- ✔ **The cervical, or neck, curvature** has 7 vertebrae, with the *atlas* and *axis* bones positioned in the first and second spots, respectively. (In a sense, the atlas bone holds the world of the head on its shoulders, as the Greek god Atlas held the Earth.)

- ✔ **The thoracic, or chest, curvature** has 12 vertebrae that articulate with the ribs, which attach to the sternum anteriorly by costal cartilage, forming the rib cage.

- ✔ **The lumbar, or small of the back, curvature** contains 5 vertebrae and carries most of the weight of the body, which means that it generally suffers the most stress.

- ✔ **The pelvic curvature** includes the 5 fused vertebrae of the *sacrum* anchoring the pelvic girdle and 4 fused vertebrae of the *coccyx*, or tailbone. (Thanks to evolution, there's no need for tail openings in our trousers.)

The spinal cord extends down the center of the vertebrae only from the base of the brain to the uppermost lumbar vertebrae.

Each vertebra consists of a body and a vertebral arch, which features a long dorsal projection called a *spinous process* that provides a point of attachment for muscles and ligaments. On either side of this are the *laminae,* broad plates of bone on the posterior surface that form a bony covering over the spinal canal. The laminae attach to the two *transverse processes,* which in turn are attached to the body of the vertebra by regions called the *pedicles.*

Articulating, or connecting, to the vertebral column are the 12 pairs of ribs that make up the thoracic cage. All 12 pairs attach to the thoracic vertebrae, but the first 7 pairs attach to the *sternum*, or breastbone, by *costal cartilage;* they're called *true ribs.* Pairs 8, 9, and 10 attach to the cartilage of the seventh pair, which is why they're called *false ribs.* The last two pairs aren't attached in front at all, so they're called *floating ribs.*

The sternum has three parts:

- ✔ **Manubrium:** The superior region that articulates with the clavicle and the first two pairs of ribs is located up top, where you can feel a notch in your chest in line with your *clavicles*, or collar bones.

- ✔ **Body:** The middle part of the sternum forms the bulk of the breastbone and has notches on the sides where it articulates with the third through seventh pairs of ribs.

- ✔ **Xiphoid process:** The lowest part of the sternum is an attachment point for the diaphragm and some abdominal muscles.

Emergency medical technicians learn to administer CPR at least three finger widths above the xiphoid.

57.–71. Use the terms that follow to identify the bones, sutures, and landmarks of the skull shown in Figure 5-2.

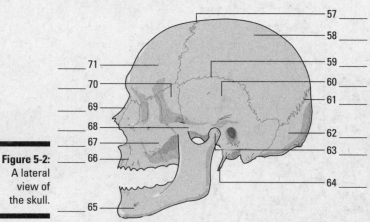

Figure 5-2: A lateral view of the skull.

LifeART Image Copyright © 2007. Wolters Kluwer Health — Lippincott Willams & Wilkins

a. Temporal bone

b. Sphenoid bone

c. Nasal bone

d. Styloid process

e. Frontal bone

f. Squamosal suture

g. Maxilla

h. Parietal bone

i. Lambdoidal suture

j. Occipital bone

k. Zygomatic bone

l. Coronal suture

m. Mandibular condyle

n. Zygomatic process

o. Mandible

72.–75. Match the skull bones with their connecting sutures.

72. _____ Frontal and parietals   a. Squamosal suture

73. _____ Occipital and parietals   b. Lambdoidal suture

74. _____ Parietal and parietal   c. Sagittal suture

75. _____ Temporal and parietal   d. Coronal suture

76.–82. Use the terms that follow to identify the bones and landmarks of the skull shown in Figure 5-3.

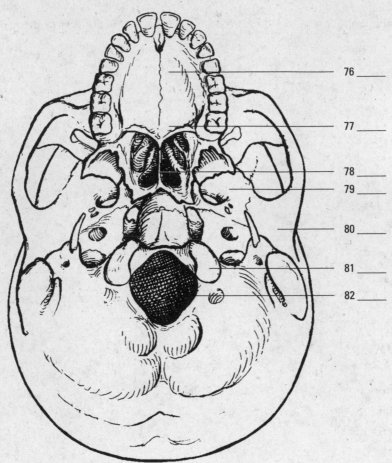

76 _____

77 _____

78 _____

79 _____

80 _____

81 _____

82 _____

Figure 5-3:
Inferior view
of the skull.

a. Vomer

b. Mandibular fossa

c. Foramen magnum

d. Palatine bone

e. Occipital condyle

f. Maxilla

g. Sphenoid bone

83.–88. Use the terms that follow to identify the bones and landmarks of the skull shown in Figure 5-4.

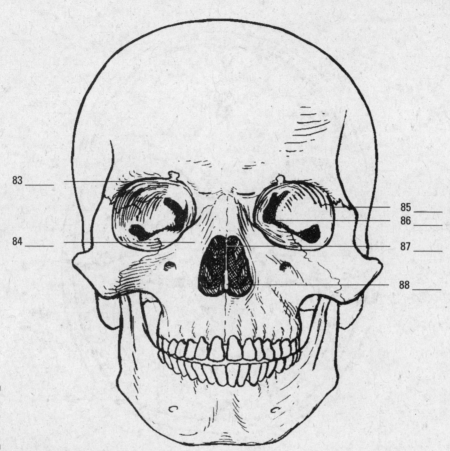

Figure 5-4:
Frontal view
of the skull.

LifeART Image Copyright © 2007. Wolters Kluwer Health — Lippincott Willams & Wilkins

a. Zygomatic bone

b. Vomer

c. Ethmoid bone

d. Sphenoid bone

e. Lacrimal bone

f. Maxilla

89.–105. Use the terms that follow to identify the bones and landmarks of the cranial cavity shown in Figure 5-5.

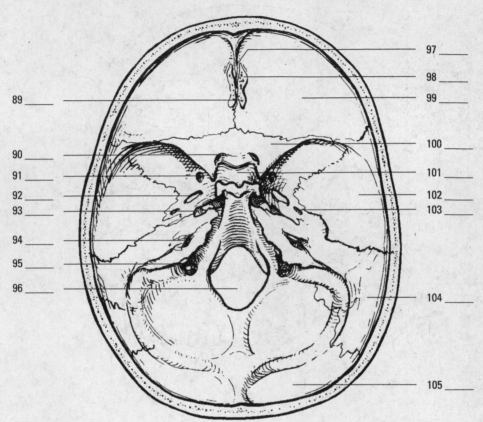

89 ___

90 ___
91 ___
92 ___
93 ___

94 ___

95 ___

96 ___

97 ___
98 ___
99 ___

100 ___

101 ___

102 ___
103 ___

104 ___

105 ___

Figure 5-5:
Cranial
cavity.

a. Internal auditory (acoustic) meatus

b. Parietal bone

c. Foramen ovale

d. Frontal bone

e. Foramen lacerum

f. Cribriform plate

g. Sella turcica

h. Temporal bone

i. Foramen spinosum

j. Occipital bone

k. Olfactory foramina

   l. Foramen rotundum

   m. Foramen magnum

   n. Crista galli

   o. Sphenoid bone

   p. Optic foramen (canal)

   q. Jugular foramen

106.–109. Use the terms that follow to identify the sinuses shown in Figure 5-6.

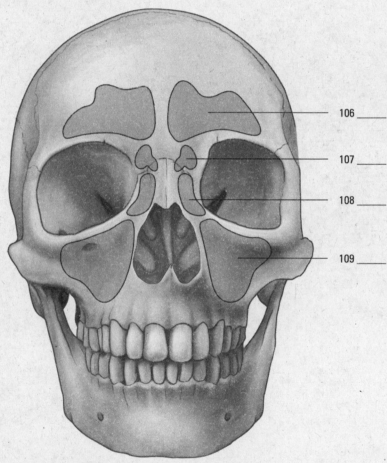

106 _____

107 _____

108 _____

109 _____

Figure 5-6:
Sinus view
of the skull.

Illustration by Imagineering Media Services Inc.

   a. Sphenoid sinus

   b. Frontal sinus

   c. Maxillary sinus

   d. Ethmoid sinus

110.–118. Use the terms that follow to identify the regions, structures, and landmarks of the vertebral column shown in Figure 5-7.

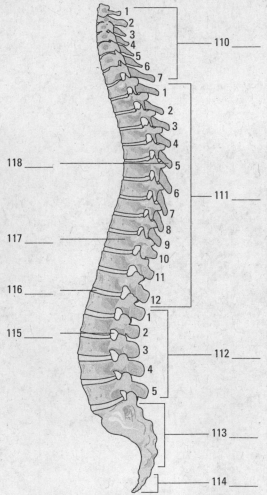

Figure 5-7: Vertebral column.

a. Coccyx

b. Intervertebral foramen

c. Thoracic vertebrae or curvature

d. Sacrum

e. A vertebra

f. Cervical vertebrae or curvature

g. Lumbar vertebrae or curvature

h. A spinous process

i. Intervertebral disc

119.–127. Use the terms that follow to identify the landmarks of the vertebra shown in Figure 5-8.

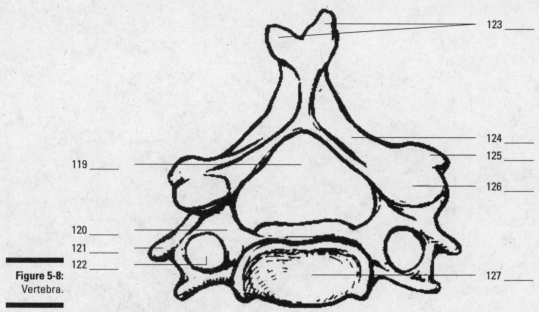

Figure 5-8:
Vertebra.

LifeART Image Copyright © 2007. Wolters Kluwer Health — Lippincott Willams & Wilkins

a. Vertebral foramen

b. Lamina

c. Transverse process

d. Spinous process (bifid)

e. Superior articulating facet

f. Body

g. Transverse foramen

h. Inferior articulating facet

i. Pedicle

128.–133. Use the terms that follow to identify the regions, landmarks, and structures of the thoracic cage shown in Figure 5-9.

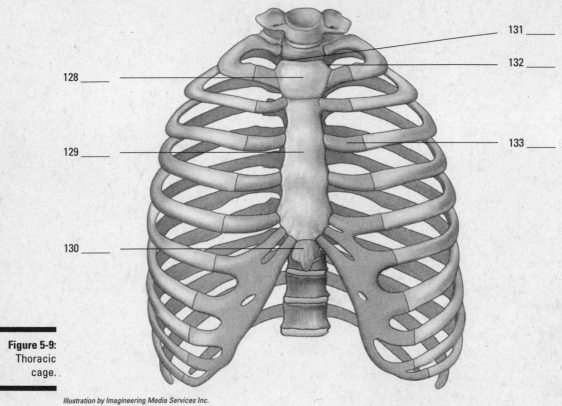

128 _____

129 _____

130 _____

131 _____

132 _____

133 _____

Figure 5-9:
Thoracic
cage.

Illustration by Imagineering Media Services Inc.

    a. Xiphoid process

    b. Jugular notch

    c. Body

    d. Costal cartilage

    e. Clavicular notch

    f. Manubrium

134.–139. Fill in the blanks to complete the following sentences:

The organs protected by the thoracic cage include the **134.** _____ and the **135.** _____. The first seven pairs of ribs attach to the sternum by the costal cartilage and are called **136.** _____ ribs. Pairs 8 through 10 attach to the costal cartilage of the seventh pair and not directly to the sternum, so they're called **137.** _____ ribs. The last two pairs, 11 and 12, are unattached anteriorly, so they're called **138.** _____ ribs. There's one bone in the entire skeleton that doesn't articulate with any other bones but nonetheless is considered part of the axial skeleton. It's called the **139.** _____ bone.

Appendicular Skeleton: Reaching Beyond Our Girdles

Whereas the axial skeleton lies along the body's central axis, the appendicular skeleton's 126 bones include those in all four appendages — arms and legs — plus the two primary girdles to which the appendages attach: the pectoral (chest) girdle and the pelvic (hip) girdle.

The pectoral girdle is made up of a pair of clavicles, or collar bones, which attach to the sternum medially and to the scapula laterally articulating with the acromion process, and a pair of scapulae, better known as shoulder blades. Each scapula has a depression in it called the *glenoid fossa* where the head of the *humerus* (upper arm bone) is attached. The lower end of the humerus articulates with the forearm's long ulna bone to form the elbow joint. The process called the *olecranon* forms the elbow and is also referred to as the funny bone, although banging it into something usually feels anything but funny. The forearm also contains a bone called the radius; together the ulna and radius articulate with the eight small carpal bones that form the wrist. The carpals articulate with the five metacarpals that form the hand, which in turn connect with the *phalanges* (finger bones), which are found as a pair in the thumb and as triplets in each of the fingers. (Toe bones also are called phalanges, but we get to that in a moment.)

The pelvic girdle consists of two hip bones, called *os coxae*, as well as the *sacrum* and *coccyx,* more commonly referred to as the tail bone. During early developmental years, the os coxa consists of three separate bones — the ilium, the ischium, and the pubis — that later fuse into one bone sometime between the ages of 16 and 20. Posteriorly, the os coxa articulates with the sacrum, forming the sacroiliac joint, the source of much lower back pain; it's formed by the connection of the hip bones at the sacrum. Toward the front of the pelvic girdle, the two os coxae join to form the *symphysis pubis,* which is made up of fibrocartilage. A cup-like socket called the *acetabulum* articulates with the ball-shaped head of the leg's *femur* (thigh bone). (The femur is the longest bone in the body.) The femur articulates with the tibia (shin bone) at the knee, which is covered by the patella (kneecap). Also inside each lower leg is the fibula bone, which joins with the tibia to connect with the seven tarsal bones that make up the ankle. The tarsals join with the five metatarsals that form the foot, which in turn connect to the phalanges of the toes — a pair of phalanges in the big toe and triplets in each of the other toes.

140.–159. Use the terms that follow to identify the bones and structures of the appendicular skeleton shown in Figure 5-10.

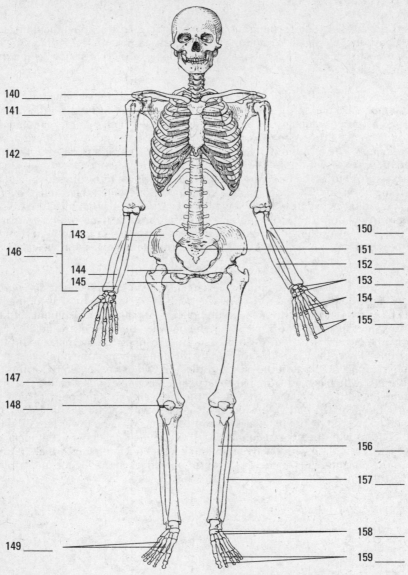

140 _____
141 _____
142 _____
143 _____
146 _____
144 _____
145 _____
147 _____
148 _____
149 _____
150 _____
151 _____
152 _____
153 _____
154 _____
155 _____
156 _____
157 _____
158 _____
159 _____

Figure 5-10:
The appendicular skeleton.

LifeART Image Copyright © 2007. Wolters Kluwer Health — Lippincott Willams & Wilkins

a. Tibia

b. Ulna

c. Scapula

d. Metatarsals

e. Carpals

    f. Phalanges of the feet

    g. Ilium

    h. Clavicle

    i. Fibula

    j. Patella

    k. Humerus

    l. Pubis

    m. Radius

    n. Os coxa

    o. Sacrum

    p. Phalanges of the hand

    q. Metacarpals

    r. Tarsals

    s. Femur

    t. Ishium

160. The structure of the scapula that articulates with the clavicle is the

    a. Acromion process

    b. Glenoid fossa

    c. Coracoid process

    d. Spine

161. The point of attachment for the biceps muscle on the radius is the

    a. Radial notch

    b. Styloid process

    c. Radial tuberosity

    d. Ulnar notch

162. The structure of the humerus that articulates with the head of the radius is the

    a. Deltoid tuberosity

    b. Olecranon fossa

    c. Trochlea

    d. Capitulum

163. The patella is what kind of bone?

    a. Wormian

    b. Sesamoid

    c. Pisiform

    d. Hallux

164. Which of these bones is not part of the pelvic girdle?

 a. Ilium

 b. Lumbar vertebrae

 c. Sacrum

 d. Ischium

165. The prominence that forms the elbow is the

 a. Olecranon process

 b. Trochlear notch

 c. Radial notch

 d. Coronoid process

166. The ulna articulates with the humerus at the

 a. Deltoid tuberosity

 b. Greater tubercle

 c. Capitulum

 d. Trochlea

167. The socket for the head of the femur is the

 a. Obturator foramen

 b. Acetabulum

 c. Ischial tuberosity

 d. Greater sciatic notch

168. The largest and strongest tarsal bone is the

 a. Talus

 b. Cuboid

 c. Navicular

 d. Calcaneus

169. A person complaining of problems in their sacroiliac has pain in the

 a. Lower back

 b. Neck

 c. Feet

 d. Hands

Arthrology: Articulating the Joints

Arthrology, which stems from the ancient Greek word *arthros* (meaning "jointed"), is the study of those structures that hold bones together, allowing them to move to varying degrees — or fixing them in place — depending on the design and function of the joint. The term *articulation*, or *joint*, applies to any union of bones, whether it moves freely or not at all.

Inside some joints, such as knees and elbows, are fluid-filled sacs called *bursae* that help reduce friction between tendons and bones; inflammation in these sacs is called *bursitis.* Some joints are stabilized by connective tissue called *ligaments* that range from bundles of collagenous fibers that restrict movement and hold a joint in place to elastic fibers that can repeatedly stretch and return to their original shapes.

The three types of joints are as follows:

- **Fibrous:** Fibrous tissue rigidly joins the bones in a form of articulation called *synarthrosis,* which is characterized by no movement at all. The sutures of the skull are fibrous joints.

- **Cartilaginous:** This type of joint is found in two forms:

  - Synchondrosis articulation involves rigid cartilage that allows no movement, such as the joint between the ribs, costal cartilage, and sternum.

  - Symphysis joints occur where cartilage fuses bones in such a way that pressure can cause slight movement, called *amphiarthrosis.* Examples include the intervertebral discs and the symphysis pubis.

- **Synovial:** Also known as *diarthrosis,* or freely moving, joints, this type of articulation involves a synovial cavity, which contains articular fluid secreted from the synovial membrane to lubricate the opposing surfaces of bone. The synovial membrane is covered by a fibrous joint capsule layer that's continuous with the periosteum of the bone. Ligaments surrounding the joint strengthen the capsule and hold the bones in place, preventing dislocation. In some synovial joints, such as the knee, fibrous connective tissue called *meniscus* develops in the cavity, dividing it into two parts. In the knee, this meniscus stabilizes the joint and acts as a shock absorber.

There are six classifications of moveable, or synovial, joints:

- **Gliding:** Curved or flat surfaces slide against one another, such as between the carpal bones in the wrist or between the tarsal bones in the ankle.

- **Hinge:** A convex surface joints with a concave surface, allowing right-angle motions in one plane, such as elbows, knees, and joints between the finger bones.

- **Pivot (or rotary):** One bone pivots or rotates around a stationary bone, such as the atlas rotating around the odontoid process at the top of the vertebral column.

- **Condyloid:** The oval head of one bone fits into a shallow depression in another, allowing the joint to move in two directions, such as the carpal-metacarpal joint at the wrist, or the tarsal-metatarsal joint at the ankle.

- **Saddle:** Each of the adjoining bones is shaped like a saddle (the technical term is *reciprocally concavo-convex*), allowing various movements, such as the carpometacarpal joint of the thumb.

- **Ball-and-socket:** The round head of one bone fits into a cup-like cavity in the other bone, allowing movement in many directions so long as the bones are neither pulled apart nor forced together, such as the shoulder joint between the humerus and scapula and the hip joints between the femur and the os coxa.

The following are the types of joint movement:

- **Flexion:** Decrease the angle between two bones

- **Extension:** Increase the angle between two bones

- **Abduction:** Movement away from the midline of the body

✔ **Adduction:** Movement toward the midline of the body

✔ **Rotation:** Turning around an axis

✔ **Pronation:** Downward or palm downward

✔ **Supination:** Upward or palm upward

✔ **Eversion:** Turning of the sole of the foot outward

✔ **Inversion:** Turning of the sole of the foot inward

✔ **Circumduction:** The forming of a cone with the arm

_170.–175._ Match the articulations with their joint types. Some joint types may be used more than once.

170. _____ Sutures of the skull   **a.** Fibrous joint

171. _____ Fluid-filled cavity   **b.** Cartilaginous joint-synchondrosis

172. _____ Knee joint   **c.** Cartilaginous joint-symphysis

173. _____ Symphysis pubis   **d.** Synovial joint

174. _____ Epiphyseal plate

175. _____ Intervertebral discs

_176.–180._ Use the terms that follow to identify the structures that form a synovial joint shown in Figure 5-11.

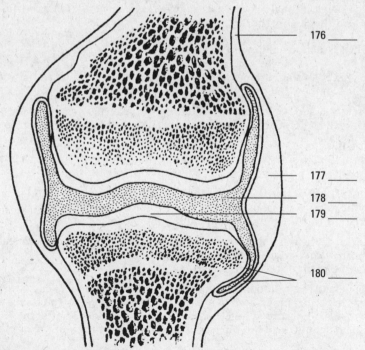

176 _____

177 _____

178 _____

179 _____

180 _____

Figure 5-11:
A synovial
joint.

LifeART Image Copyright © 2007. Wolters Kluwer Health — Lippincott Willams & Wilkins

a. Synovial (joint) cavity

b. Periosteum

c. Synovial membrane

    d. Articular cartilage

    e. Fibrous capsule

181. An immovable joint is

    a. Amphiarthrosis

    b. Synarthrosis

    c. Diarthrosis

    d. Synchondrosis

182. A freely moving joint is

    a. Amphiarthrosis

    b. Synarthrosis

    c. Diarthrosis

    d. Synchondrosis

183. The material or structure that allows for free movement in a joint is

    a. Bursa

    b. Periosteum

    c. Synovial fluid

    d. Bone marrow

184. An example of a ball-and-socket joint is the

    a. Symphysis pubis

    b. Hip

    c. Ankle

    d. Elbow

185. An example of a pivotal joint is

    a. Between the radius and the ulna

    b. The interphalanges joints

    c. Between the mandible and the temporal bone

    d. Between the tibia and the fibula

186. A saddle joint is located in

    a. The radius and the carpals

    b. The carpometacarpal joint of the thumb

    c. The occipital condyles and the atlas

    d. The metatarsophalanges joint

187. A shoulder joint ligament is the

    a. Coracohumeral ligament

    b. Popliteal ligament

    c. Ischiofemoral ligament

    d. Ligamentum arteriosum

188. A knee joint ligament is the

    a. Iliofemoral ligament

    b. Coracohumeral ligament

    c. Oblique popliteal ligament

    d. Annular ligament

189. A hip joint ligament is the

    a. Subscapularis ligament

    b. Pubofemoral ligament

    c. Glenohumeral ligament

    d. Coracohumeral ligament

190. The structure in the knee that divides the synovial joint into two separate compartments is the

    a. Bursa

    b. Joint fat

    c. Tendon sheath

    d. Meniscus or articular disc

191.–200. Match the type of joint movement with its description.

191. _____ Flexion	**a.** Upward or palm upward
192. _____ Extension	**b.** Decrease the angle between two bones
193. _____ Abduction	**c.** Turning of the sole of the foot inward
194. _____ Adduction	**d.** Downward or palm downward
195. _____ Rotation	**e.** Increase the angle between two bones
196. _____ Pronation	**f.** Turning of the sole of the foot outward
197. _____ Supination	**g.** Movement away from the midline of the body
198. _____ Eversion	**h.** The forming of a cone with the arm
199. _____ Inversion	**i.** Turning around an axis
200. _____ Circumduction	**j.** Movement toward the midline of the body

Answers to Questions on the Skeleton

The following are answers to the practice questions presented in this chapter.

1   The formation of red blood cells by the bone marrow is known as **b. hematopoiesis.** The term hemopoiesis also would be correct here, but it's not one of the answer options.

2   The following mineral is stored inside bones for later use: **d. All of the above (phosphorous, calcium, and magnesium).** The bones act as a mineral bank for the entire body.

3   Besides support and protection, the skeleton serves other important functions, including **b. locomotion.** You'd be a motionless blob without a skeleton to support coordinated movement.

4   The curvatures in some bone structures serve the following purpose: **a. Support the body.** And now you know why your back aches most where it curves.

5   The most abundant mineral in the body is **c. calcium.**

6   A decrease in calcium concentration in the body fluids by 35 percent causes **c. convulsions.** Go much beyond that and the situation becomes fatal.

7   Calcitonin is produced by the **a. thyroid gland.** There's that busy little thyroid gland, controlling metabolism.

8   Bones are encased by a membrane called the **b. periosteum.** Back to Greek again: *peri* means "around" and *osteon* means "bone," so the periosteum is "around the bone."

9   A bone's encapsulating membrane is attached by **c. Sharpey's fibers.** Described by anatomist William Sharpey in 1846, these are also called *perforating fibers*.

10   Bone marrow can be found inside the **a. medullary cavity.** This is where you'll find yellow marrow, although in infants red marrow also is present.

11   The six structures in the skull of an infant are called **b. fontanels.** These separate floating plates are why you can see a bald baby's pulse throbbing on the top of its head.

12   Volkmann's canals **a. are found in cancellous bone only.** Ironically, anatomist Alfred Wilhelm Volkmann was most noted for his observations of the physiology of the nervous system, not bones.

13   Blood vessels entering through Volkmann's canals reach the bone cells through the **b. Haversian canal.** The name comes from the first physician to describe them, Clopton Havers.

14   The hormone released when calcium ion concentration is abnormally high is **b. calcitonin.** This peptide hormone lowers plasma calcium.

15 – 25   Bones are first laid down as **15. cartilage** during the fifth week after conception. Development of the bone begins with **16. calcification**, the depositing of calcium and phosphorus. Next, the blood supply entering the cartilage brings **17. osteoblasts** that attach themselves to the cartilage. Ossification in long bones begins in the **18. diaphysis** of the long bone and moves toward the **19. epiphysis** of the bone. The epiphyseal and diaphyseal areas remain separated by a layer of uncalcified cartilage called the **20. epiphyseal plate.**

Another very large cell that enters with the blood supply is the **21. osteoclast**, which helps absorb the cartilage as ossification occurs. Later it helps absorb bone tissue from the center of the long bone's shaft, forming the **22. medullary or marrow** cavity. After ossification, the spaces that were formed by the osteoclasts join together to form **23. Haversian canal** systems, which contain the blood vessels, lymphatic vessels, and nerves. Unlike bones in the rest of the body, those of the skull and mandible (lower jaw) are first laid down as **24. membrane**. In the skull, the edges of the bone don't ossify in the fetus but remain membranous and form **25. fontanels**.

26 Vertebrae of the vertebral column: **b. Irregular bone**

27 Femur in thigh: **c. Long bone**

28 Sternum: **a. Flat bone**

29 Tarsals in ankle: **d. Short bone**

30 Humerus in upper arm: **c. Long bone**

31 Phalanges in fingers and toes: **d. Short bone**

32 Scapulae of shoulder: **a. Flat bone**

33 Kneecap: **b. Irregular bone**

34 Carpals in wrist: **d. Short bone**

35 An abrupt or pointed projection: **i. Spine**

36 A large, usually blunt process: **k. Trochanter**

37 A designation for any prominence or prolongation: **h. Process**

38 A large, often rough eminence: **m. Tuberosity**

39 A prominent ridge: **b. Crest**

40 A large, rounded articular end of a bone; often set off from the shaft by the neck: **f. Head**

41 An oval articular prominence of a bone: **a. Condyle**

42 A smooth, flat or nearly flat articulating surface: **c. Facet**

43 A deeper depression: **e. Fossa**

44 A groove: **j. Sulcus**

45 A hole: **d. Foramen**

46 A canal or opening to a canal: **g. Meatus**

47 A smaller, rounded eminence: **l. Tubercle**

48–56 Following is how Figure 5-1, the long bone, should be labeled.

48. **h. Articulating cartilage**; 49. **d. Spongy bone tissue**; 50. **g. Red bone marrow**; 51. **i. Compact bone tissue**; 52. **e. Medullary or nutrient artery**; 53. **b. Medullary cavity**; 54. **f. Proximal epiphysis**; 55. **a. Diaphysis**; 56. **c. Distal epiphysis**

57—71 Following is how Figure 5-2, the lateral view of the skull, should be labeled.

> 57. **l. Coronal suture**; 58. **h. Parietal bone**; 59. **f. Squamosal suture**; 60. **a. Temporal bone**; 61. **i. Lamdoidal suture**; 62. **j. Occipital bone**; 63. **m. Mandibular condyle**; 64. **d. Styloid process**; 65. **o. Mandible**; 66. **g. Maxilla**; 67. **k. Zygomatic bone**; 68. **n. Zygomatic process**; 69. **c. Nasal bone**; 70. **b. Sphenoid bone**; 71. **e. Frontal bone**

72 Frontal and parietals: **d. Coronal suture**

73 Occipital and parietals: **b. Lambdoidal suture**

74 Parietal and parietal: **c. Sagittal suture**

75 Temporal and parietal: **a. Squamosal suture**

76—82 Following is how Figure 5-3, the inferior view of the skull, should be labeled.

> 76. **f. Maxilla**; 77. **d. Palatine bone**; 78. **a. Vomer**; 79. **g. Sphenoid bone**; 80. **b. Mandibular fossa**; 81. **e. Occipital condyle**; 82. **c. Foramen magnum**

83—88 Following is how Figure 5-4, the frontal view of the skull, should be labeled.

> 83. **d. Sphenoid bone**; 84. **f. Maxilla**; 85. **a. Zygomatic bone**; 86. **e. Lacrimal bone**; 87. **c. Ethmoid bone**; 88. **b. Vomer**

89—105 Following is how Figure 5-5, the cranial cavity, should be labeled.

> 89. **f. Cribriform plate**; 90. **p. Optic foramen (canal)**; 91. **l. Foramen rotundum**; 92. **c. Foramen ovale**; 93. **e. Foramen lacerum**; 94. **a. Internal auditory (acoustic) meatus**; 95. **q. Jugular foramen**; 96. **m. Foramen magnum**; 97. **n. Crista galli**; 98. **k. Olfactory foramina**; 99. **d. Frontal bone**; 100. **o. Sphenoid bone**; 101. **g. Sella turcica**; 102. **h. Temporal bone**; 103. **i. Foramen spinosum**; 104. **b. Parietal bone**; 105. **j. Occipital bone**

106—109 Following is how Figure 5-6, the sinus view of the skull, should be labeled.

> 106. **b. Frontal sinus**; 107. **d. Ethmoid sinus**; 108. **a. Sphenoid sinus**; 109. **c. Maxillary sinus**

110—118 Following is how Figure 5-7, the vertebral column, should be labeled.

> 110. **f. Cervical vertebrae or curvature**; 111. **c. Thoracic vertebrae or curvature**; 112. **g. Lumbar vertebrae or curvature**; 113. **d. Sacrum**; 114. **a. Coccyx**; 115. **b. Intervertebral foramen**; 116. **i. Intervertebral disc**; 117. **e. A vertebra**; 118. **h. A spinous process**

119—127 Following is how Figure 5-8, the vertebra, should be labeled.

> 119. **a. Vertebral foramen**; 120. **i. Pedicle**; 121. **c. Transverse process**; 122. **g. Transverse foramen**; 123. **d. Spinous process (bifid)**; 124. **b. Lamina**; 125. **h. Inferior articulating facet**; 126. **e. Superior articulating facet**; 127. **f. Body**

128—133 Following is how Figure 5-9, the thoracic cage, should be labeled.

> 128. **f. Manubrium**; 129. **c. Body**; 130. **a. Xiphoid process**; 131. **b. Jugular notch**; 132. **e. Clavicular notch**; 133. **d. Costal cartilage**

134—139 The organs protected by the thoracic cage include the **134. heart** and the **135. lungs**. The first seven pairs of ribs attach to the sternum by the costal cartilage and are called **136. true** ribs. Pairs 8 through 10 attach to the costal cartilage of the seventh pair and not directly to the sternum, so they're called **137. false** ribs. The last two pairs, 11 and 12, are unattached anteriorly, so they're called **138. floating** ribs. There's one bone in the entire skeleton that doesn't articulate with any other bones but nonetheless is considered part of the axial skeleton. It's called the **139. hyoid** bone.

140–159 Following is how Figure 5-10, the frontal view of the skeleton, should be labeled.

140. **h. Clavicle**; 141. **c. Scapula**; 142. **k. Humerus**; 143. **g. Ilium**; 144. **l. Pubis**; 145. **t. Ishium**; 146. **n. Os coxa**; 147. **s. Femur**; 148. **j. Patella**; 149. **d. Metatarsals**; 150. **m. Radius**; 151. **o. Sacrum**; 152. **b. Ulna**; 153. **e. Carpals**; 154. **q. Metacarpals**; 155. **p. Phalanges of the hand**; 156. **a. Tibia**; 157. **i. Fibula**; 158. **r. Tarsals**; 159. **f. Phalanges of the feet**

160 The structure of the scapula that articulates with the clavicle is the **a. acromion process.**

161 The point of attachment for the biceps muscle on the radius is the **c. radial tuberosity.**

162 The structure of the humerus that articulates with the head of the radius is the **d. capitulum.**

163 The patella is what kind of bone? **b. Sesamoid**

164 Which of these bones is not part of the pelvic girdle? **b. Lumbar vertebrae**

165 The prominence that forms the elbow is the **a. olecranon process.**

166 The ulna articulates with the humerus at the **d. trochlea.**

167 The socket for the head of the femur is the **b. acetabulum.**

168 The largest and strongest tarsal bone is the **d. calcaneus.**

169 A person complaining of problems in their sacroiliac has pain in the **a. lower back.**

170 Sutures of the skull: **a. Fibrous joint**

171 Fluid-filled cavity: **d. Synovial joint**

172 Knee joint: **d. Synovial joint**

173 Symphysis pubis: **c. Cartilaginous joint-symphysis**

174 Epiphyseal plate: **b. Cartilaginous joint-synchondrosis**

175 Intervertebral discs: **c. Cartilaginous joint-symphysis**

176–180 Following is how Figure 5-11, the synovial joint, should be labeled.

176. **b. Periosteum**; 177. **e. Fibrous capsule**; 178. **a. Synovial (joint) cavity**; 179. **d. Articular cartilage**; 180. **c. Synovial membrane**

181 An immovable joint is **b. synarthrosis.**

182 A freely moving joint is **c. diarthrosis.**

183 The material or structure that allows for free movement in a joint is **c. synovial fluid.**

184 An example of a ball-and-socket joint is the **b. hip.**

185 An example of a pivotal joint is **a. between the radius and the ulna.**

186 A saddle joint is located in **b. The carpometacarpal joint of the thumb.**

187 A shoulder joint ligament is **a. The coracohumeral ligament.**

188 A knee joint ligament is the **c. oblique popliteal ligament.**

189 A hip joint ligament is the **b. pubofermoral ligament.**

190 The structure in the knee that divides the synovial joint into two separate compartments is the **d. meniscus or articular disc.**

191 Flexion: **b. Decrease the angle between two bones**

192 Extension: **e. Increase the angle between two bones**

193 Abduction: **g. Movement away from the midline of the body**

194 Adduction: **j. Movement toward the midline of the body**

195 Rotation: **i. Turning around an axis**

196 Pronation: **d. Downward or palm downward**

197 Supination: **a. Upward or palm upward**

198 Eversion: **f. Turning of the sole of the foot outward**

199 Inversion: **c. Turning of the sole of the foot inward**

200 Circumduction: **h. The forming of a cone with the arm**

Chapter 6

Getting in Gear: The Muscles

. . . .

In This Chapter

▶ Understanding the functions and structure of muscles

▶ Classifying types of muscle

▶ Pulling together: Muscles as organs

▶ Breaking down muscle contractions, tone, and power

▶ Deciphering muscle names

. . . .

Much of what we think of as "the body" centers around our muscles and what they can do, what we want them to do, and how tired we get trying to make them do it. With all that muscles do and are, it's hard to believe the word "muscle" is rooted in the Latin word *musculus,* which is a diminutive of the word for "mouse." Well, the muscle is a mouse that roars. Muscles make up most of the fleshy parts of the body and average 43 percent of the body's weight. Layered over the skeleton, they largely determine the body's form. There are over 500 muscles large enough to be seen by the unaided eye, and thousands more are visible only through a microscope. Although there are three distinct types of muscle tissue, every muscle in the human body shares one important characteristic: *contractility,* the ability to shorten, or contract.

Flexing Your Muscle Knowledge

The study of muscles is called *myology* after the Greek word *mys,* which means "mouse." Muscles perform a number of functions vital to maintaining life, including

✔ **Movement:** Skeletal muscles (those attached to bones) convert chemical energy into mechanical work, producing movement ranging from finger tapping to a swift kick of a ball by contracting, or shortening. Reflex muscle reactions protect your fingers when you put them too close to a fire and startle you into watchfulness when an unexpected noise sounds. Many purposeful movements require several sets, or groups, of muscles to work in unison.

✔ **Vital functions:** Without muscle activity, you die. Muscles are doing their job when your heart beats, when your blood vessels constrict, and when your intestines squeeze food along your digestive tract in *peristalsis.*

✔ **Antigravity:** Perhaps that's overstating it, but muscles do make it possible for you to stand and move about in spite of gravity's ceaseless pull. Did your mother tell you to improve your posture? Just think how bad it would be without any muscles!

✔ **Heat generation:** You shiver when you're cold and stamp your feet and jog in place when you need to warm up. That's because chemical reactions in muscles result in heat, helping to maintain the body's temperature.

✔ **Keep the body together:** Muscles are the warp and woof of your body's structure, binding one part to another.

As you may remember from studying tissues, muscle cells — called *fibers* — are some of the longest in the body. Fibers are held together by connective tissue and enclosed in a fibrous sheath called *fascia*. Some muscle fibers contract rapidly, whereas others move at a leisurely pace. Generally speaking, however, the smaller the structure to be moved, the faster the muscle action. Exercise can increase the thickness of muscle fibers, but it doesn't make new fibers. Skeletal muscles have a rich vascular supply that dilates during exercise to give the working muscle the extra oxygen it needs to keep going.

Two processes are central to muscle development in the developing embryo: *myogenesis*, during which muscle tissue is formed; and *morphogenesis*, when the muscles form into internal organs. By the eighth week of gestation, a fetus is capable of coordinated movement.

Following are some important muscle terms to know:

- **Fascia:** Loose, or areolar, connective tissue that holds muscle fibers together to form a muscle organ

- **Fiber:** An individual muscle cell

- **Insertion:** The more movable attachment of a muscle

- **Ligament:** Elastic connective tissue that supports joints and anchors organs

- **Motor nerve:** Nerve that stimulates contraction of a muscle

- **Myofibril:** Fibrils within a muscle cell that contain protein filaments such as actin and myosin that slide during contraction, shortening the fiber (or cell)

- **Origin:** The immovable attachment of a muscle, or the point at which a muscle is anchored by a tendon to the bone

- **Sarcoplasm:** The cellular cytoplasm in a muscle fiber

- **Tendon:** Connective tissue made up of collagen, a fibrous protein that attaches muscles to bone; lets muscles apply their force at some distance from where a contraction actually takes place

- **Tone, or tonus:** State of tension present to a degree at all times, even when the muscle is at rest

Complete the following practice questions to see how well you understand the basics of myology:

1. Which of the following is *not* a true statement?

   a. Muscles represent 90 percent of the total body weight.

   b. The ancient Greek word *mys* means "mouse."

   c. The muscles covering the bones largely determine the form of the body.

   d. Posture is an expression of muscle action.

2. Muscle functions include

   a. Support of the bony tissues of the body

   b. Blood formation

   c. Converting chemical energy into mechanical work

   d. Only a and c

3. A necessary property for a muscle to perform work is
   a. Extensibility
   b. Contractility
   c. Elasticity
   d. All of the above

4. The cellular unit in muscle tissue is the
   a. Filament
   b. Myofibril
   c. Fiber
   d. Fasciculus

5. A partial state of contraction, in part, defines
   a. Rigor
   b. Tonus
   c. Clovus
   d. Paralysis

6. It's possible to completely relax every muscle in the body.
   a. True
   b. False

7. During embryonic development, tissue development is called
   a. Myogelosis
   b. Morphogenesis
   c. Myogenesis
   d. Morpholysis

8. Exercise forms new muscle fibers.
   a. True
   b. False

Classifications: Smooth, Cardiac, and Skeletal

Muscle tissue is classified in three ways based on the tissue's function, shape, and structure:

> ✔ **Smooth muscle tissue:** So-called because it doesn't have the cross-striations typical of other kinds of muscle, the spindle-shaped fibers of smooth muscle tissue do have faint longitudinal striping. This muscle tissue forms into sheets and makes up the walls of hollow organs such as the stomach, intestines, and bladder. The tissue's involuntary movements are relatively slow, so contractions last

longer than those of other muscle tissue, and fatigue is rare. Each fiber is about 6 microns in diameter and can vary from 15 microns to 500 microns long. If arranged in a circle inside an organ, contraction constricts the cavity inside the organ. If arranged lengthwise, contraction of smooth muscle tissue shortens the organ.

✔ **Cardiac muscle tissue:** Found only in the heart, cardiac muscle fibers are branched, cross-striated, feature one central nucleus, and move through involuntary control. An electron microscope view of the tissue shows separate fibers tightly pressed against each other, forming cellular junctions called *intercalated discs* that look like tiny, dark-colored plates. Some experts believe intercalated discs are not cellular junctions but rather special structures that help move an electrical impulse throughout the heart.

✔ **Skeletal muscle tissue:** This is the tissue that most people think of as muscle. It's the only muscle subject to voluntary control through the central nervous system. Its long, striated cylindrical fibers contract quickly but tire just as fast. Skeletal muscle, which is also what's considered meat in animals, is 20 percent protein, 75 percent water, and 5 percent organic and inorganic materials. Each multinucleated fiber is encased in a thin, transparent membrane called a *sarcolemma* that receives and conducts stimuli. The fibers, which vary from 10 microns to 100 microns in diameter and up to 4 centimeters in length, are subdivided lengthwise into tiny myofibrils roughly 1 micron in diameter that are suspended in the cell's sarcoplasm.

The following practice questions test your knowledge of muscle classifications:

9. This type of muscle tissue lacks cross-striations.

   a. Cardiac

   b. Smooth

   c. Skeletal

   d. Contracting

10. Skeletal muscle fibers are encased in

   a. A sarcolemma

   b. Sarcoplasm

   c. Sarcomeres

   d. A sarcophagus

11. Which muscle type appears only in a single organ?

   a. Contractile

   b. Smooth

   c. Cardiac

   d. Skeletal

12. Intercalated discs

   a. Anchor cardiac muscle fibers to one another

   b. May play a role in moving electrical impulses through the heart

   c. Are found only in the muscles of the back

   d. Contribute to tactile perception

Contracting for a Contraction

Before we can explain how muscles do what they do, it's important that you understand the anatomy of how they're put together. Use Figure 6-1 as a visual guide as you read through this section.

We base this description of muscle on the most studied classification of muscle: skeletal. Each fiber packed inside the sarcolemma contains hundreds, or even thousands, of myofibril strands made up of alternating filaments of the proteins *actin* and *myosin*. Actin and myosin are what give skeletal muscles their striated appearance, with alternating dark and light bands. The dark bands are called *anisotropic*, or *A-bands*. The light bands are called *isotropic*, or *I-bands*. In the center of each I-band is a line called the *Z-line* that divides the myofibril into smaller units called *sarcomeres*. At the center of the A-band is a less-dense region called the *H-zone*.

Now, here's where the actin and myosin come in. Each sarcomere contains thick filaments of myosin in the A-band and thin filaments of actin primarily in the I-band but extending a short distance between the myosin filaments into the A-band. Actin filaments don't extend all the way into the central area of the A-band, which explains why the less-dense H-zone can be found there. Those thin actin filaments are anchored to the Z-line at their midpoints, which holds them in place and creates a structure against which the filaments exert their pull during contraction.

The theory of contraction called the *Interdigitating Filament Model of Muscle Contraction*, or the *Sliding Theory of Muscle Contraction*, says that the myosin of the thick filaments combines with the actin of the thin filaments, forming *actomyosin* and prompting the filaments to slide past each other. As they do so, the H-zone is reduced or obliterated, pulling the Z-lines closer together and reducing the I-bands. (The A-bands don't change.) Voila! Contraction has occurred!

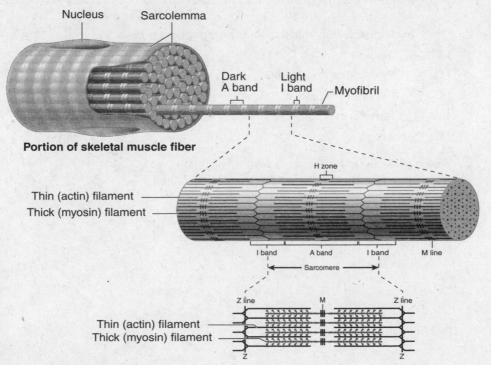

Figure 6-1:
Microscopic
anatomy of
a skeletal
muscle
fiber.

Illustration by Imagineering Media Services Inc.

So you know how muscles contract. Now you need to figure out what stimulates them to do so. We cover the details of the nervous system in Chapter 15, but here you can find out what's happening as an impulse stimulates a skeletal muscle.

The *impulse,* or *stimulus,* from the central nervous system is brought to the muscle through a nerve called the *motor,* or *efferent,* nerve. On entering the muscle, the motor nerve fibers separate to distribute themselves among the thousands of muscle fibers. Because the muscle has more fibers than the motor nerve, individual nerve fibers branch repeatedly so that a single nerve fiber innervates from 5 to as many as 200 muscle fibers. These small terminal branches penetrate the sarcolemma and form a special structure known as the *motor end plate,* or *synapse.* This neuromuscular unit consisting of one motor neuron and all the muscle fibers that it innervates is called the *motor unit.*

Interference — either chemical or physical — with the nerve pathway can affect the action of the muscle or stop the action altogether, resulting in muscle paralysis. There also are *afferent,* or *sensory,* nerves that carry information about muscle condition to the brain.

When an impulse moves through the synapse and the motor unit, it must arrive virtually simultaneously at each of the individual sarcomeres to create an efficient contraction. Enter the *transverse system,* or *T-system,* of tubules. The fiber's membrane forms deep invaginations, or inward-folding sheaths, at the Z-line of the myofibrils. The resulting inward-reaching tubules ensure that the sarcomeres are stimulated at nearly the same time.

Does it matter whether the signal received is strong or weak? Nope. That's the *all-or-none law* of muscle contraction. The fiber either contracts completely or not at all. In other words, if a single muscle fiber is going to contract, it's going to do so to its fullest extent.

Following are some practice questions that deal with muscle anatomy and contraction:

13.–17. Match each muscle component with the appropriate region.

13. _____ Myosin

14. _____ Segment of fibril from Z-line to Z-line

15. _____ Less-dense region of the A-band

16. _____ Structure to which filaments are attached

17. _____ Actin

a. H-zone

b. Z-line

c. I-band

d. A-band

e. Sarcomere

18. Which of these terms doesn't belong in the following list?

a. Anisotropic

b. Actin

c. Myosin

d. Isotropic

e. Sarcolemma

19. This part of a muscle doesn't change during contraction:

a. The H-zone

b. The A-bands

c. The I-bands

d. The Z-lines

20. A weak stimulus causes a muscle fiber to contract only partway.

   a. True

   b. False

Pulling Together: Muscles as Organs

A muscle organ has two parts:

- ✔ **The belly,** composed predominantly of muscle fibers
- ✔ **The tendon,** composed of fibrous, or collagenous, regular connective tissue. If the tendon is a flat, sheet-like structure attaching a wide muscle, it's called an *aponeurosis.*

Each muscle fiber outside of the sarcolemma is surrounded by areolar connective tissue called *endomysium* that binds the fibers together into bundles called *fasciculi* (see Figure 6-2). Each bundle, or *fasciculus,* is surrounded by areolar connective tissue called *perimysium.* All the fasciculi together make up the belly of the muscle, which is surrounded by areolar connective tissue called the *epimysium.* Blood vessels, lymph vessels, and nerves pass into the fasciculus through areolar connective tissue called the *trabecula.* These blood vessels in turn branch off into capillaries that surround the muscle fibers in the endomysium.

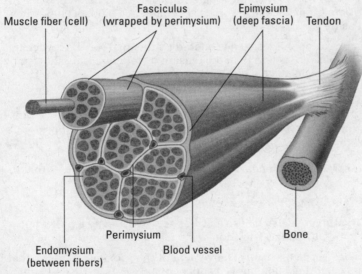

Figure 6-2:
Connective
tissue in a
muscle.

Muscle fiber (cell) Fasciculus (wrapped by perimysium) Epimysium (deep fascia) Tendon

Perimysium Bone

Endomysium (between fibers) Blood vessel

Illustration by Imagineering Media Services Inc.

21.–25. Match the muscle structures with their descriptions.

   21. _____ Membrane covering a muscle fiber

   22. _____ Bundles of muscle fibers

   23. _____ Connective tissue that surrounds a bundle of muscle fibers

   24. _____ Connective tissue through which arteries and veins enter muscle bundles

   25. _____ Flat, sheet-like tendon that serves as insertion for a large flat muscle

   a. Perimysium

   b. Aponeurosis

   c. Trabecula

   d. Fasciculi

   e. Sarcolemma

Assuming the Right Tone

As we note earlier in this chapter, when it comes to contraction of a muscle fiber, it's an all-or-nothing affair. Nonetheless, it has been demonstrated that fewer action potentials — a weaker stimulus, as it were — causes fewer motor units to become involved in a contraction. Maximum stimulus, on the other hand, brings all motor units to bear together. So it's true that a muscle organ can have varying degrees of contraction depending on the level of stimulation. As for how this can be so, one theory proposes that individual fibers have specific thresholds of excitation; thus, those with higher thresholds only respond to stronger stimuli. The other theory holds that the deeper a fiber is buried in the muscle, the less accessible it is to incoming stimuli.

In physiology, a muscle contraction is referred to as a *muscle twitch*. A twitch is the fundamental unit of recordable muscular activity. Complete fatigue occurs when no more twitches can be elicited, even with increasing intensity of stimulation.

The short lapse of time between the application of a stimulus and the beginning of muscular response is called the *latent period*. In mammalian muscle, latency is about .001 second, or one one-thousandth of a second.

Two types of muscle contraction relate to tone:

- ✔ **Isometric:** Occurs when a contracting muscle is unable to move a load (or heft a piece of luggage or push a building to one side). It retains its original length but develops *tension*. No mechanical work is accomplished, and all energy involved is expended as heat.

- ✔ **Isotonic:** Occurs when the resistance offered by the load (or the gardening hoe or the cold can of soda) is less than the tension developed, thus shortening the muscle and resulting in mechanical work.

But muscles aren't independent sole proprietors. Each muscle depends upon companions in a muscle group to assist in executing a particular movement. That's why muscles are categorized by their actions. The brain coordinates the following groups through the cerebellum.

- ✔ **Prime movers:** Just as it sounds, these muscles are the workhorses that produce movement.

- ✔ **Antagonists:** These muscles exist in opposition to prime movers.

- ✔ **Fixators or fixation muscles:** These muscles serve to steady a part while other muscles execute movement. They don't actually take part in the movement itself.

- ✔ **Synergists:** These muscles control movement of the proximal joints so that the prime movers can bring about movements of distal joints.

Flex your knowledge of muscle tone and function with these practice questions:

Q. Muscle movement that lifts an object involves an action known as

    a. Isometric

    b. Eccentric

    c. Isotonic

A. The correct answer is isotonic. When the tension leads to movement (actual work), it's isotonic.

26. Muscles that tend to counteract or slow an action are called

   a. Antagonists

   b. Fixators

   c. Primary movers

   d. Synergists

27. Which of the following statements finishes this sentence and makes it *not* true: A contracting muscle unable to move a load

   a. Involves an action known as isometric

   b. Expends energy as heat

   c. Is exemplified in the effect of the force of gravity on muscle contraction

   d. Does no mechanical work and therefore doesn't develop any tension

   e. Retains its original length

28. A muscle contraction is referred to as

   a. Latency

   b. Synergy

   c. A twitch

   d. Isotonic motion

Leveraging Muscular Power

Skeletal muscle power is nothing without lever action. The bone acts as a rigid bar, the joint is the fulcrum, and the muscle applies the force. Levers are divided into the *weight arm,* the area between the fulcrum and the weight; and the *power arm,* the area between the fulcrum and the force. When the power arm is longer than the weight arm, less force is required to lift the weight, but range, or distance, and speed are sacrificed. When the weight arm is longer, the range of action and speed increase, but power is sacrificed. Therefore, 90 degrees is the optimum angle for a muscle to attach to a bone and apply the greatest force.

Three classes of levers are at work in the body:

   ✔ **Class I, or seesaw:** The fulcrum is located between the weight and the force being applied. An example is a nod of the head: The head-neck joint is the fulcrum, the head is the weight, and the muscles in the back of the neck apply the force.

   ✔ **Class II, or wheelbarrow:** The weight is located between the fulcrum and the point at which the force is applied. An example is standing on your tiptoes: The fulcrum is the joint between the toes and the foot, the weight is the body, and the muscles in the back of the leg at the heel bone apply the force.

   ✔ **Class III:** The force is located between the weight and the fulcrum. An example is flexing your arm and showing off your biceps: The elbow joint is the fulcrum, the weight is the lower arm and hand, and the biceps insertion on the lower arm applies the force.

The direction in which the muscle fibers run also plays a critical role in leverage. Here are the possible directions:

- **Longitudinal:** Fibers run parallel to each other, or longitudinally, the length of the muscle. Example: sartorius.

- **Pennate:** Fibers attach to the sides of the tendon, which extends the length of the muscle. These come in subcategories:

    - **Unipennate,** where fibers attach to one side of the tendon; example: tibialis posterior

    - **Bipennate,** where fibers attach to two sides of the tendon; example: rectus femoris

    - **Multipennate,** where fibers attach to many sides of the tendon; example: deltoideus

- **Radiate:** Fibers converge from a broad area into a common point. Example: pectoralis major.

- **Sphincter:** Fibers are arranged in a circle around an opening. Example: orbicularis oculi.

The three types of *fasciae,* which *Gray's Anatomy* describes as "dissectable, fibrous connective tissues of the body," are as follows:

- **Superficial fasciae:** Found under the skin and consisting of two layers: an outer layer called the *panniculus adiposus* containing fat; and an inner layer made up of a thin, membranous, and highly elastic layer. Between the two layers are the superficial arteries, veins, nerves, and mammary glands.

- **Deep fasciae:** Holds muscles or structures together or separates them into groups that function in unison. It's a system of splitting, rejoining, and fusing membranes involving

    - An outer investing layer that's found under the superficial fasciae covering a large part of the body

    - An internal investing layer that lines the inside of the body wall in the torso, or trunk, region

    - An intermediate investing layer that connects the outer investing layer and the internal investing layer

- **Subserous fasciae:** Located between the internal investing layer of the deep fasciae and the peritoneum. It's the serous membrane that lines the *abdominopelvic* cavity, also known as the *peritoneal cavity.*

Got all that? Then try your hand at the following questions:

29. Which of the following in Figure 6-3 is a Class II lever?

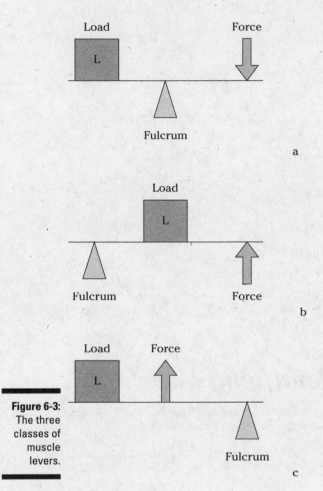

Figure 6-3:
The three classes of muscle levers.

30. Which of the following would provide the force in a Class III lever?

   a. Biceps brachii

   b. Spenius capitus

   c. Triceps brachii

   d. Gastrocnemius

31. Which of the following would produce a wide range of movement with speed while sacrificing power?

   a. Power arm and weight arm of equal lengths

   b. Long weight arm, short power arm

   c. Long power arm, shorter weight arm

32. Identify the bipennate bundle arrangement.

   a. Sartorius

   b. Rectus femoris

   c. Pectoralis major

   d. Tibialis posterior

   e. Deltoideus

33. Which of the following is considered a dissectable connective tissue?

   a. Aponeurosis

   b. Bursae

   c. Fasciae

   d. Tendons

   e. Ligaments

34. The most extensive fascia in the body is

   a. Superficial

   b. Deep

   c. Subserous

   d. None is more extensive than the other

What's In a Name? Identifying Muscles

It may seem like a jumble of meaningless Latin at first, but muscle names follow a strict convention that lets them be named for one or more of four things:

✔ **Function:** These muscle names usually have a verb root and end in a suffix (*–or* or *–eus*), followed by the name of the affected structure. Example: levator scapulae (elevates the scapulae).

✔ **Compounding points of attachment:** These muscle names blend the origin and insertion attachment with an adjective suffix (*–eus* or *–is*). Examples: sternocleidomastoideus (sternum, clavicle, and mastoid process) and sternohyoideus (sternum and hyoid).

✔ **Shape or position:** These muscle names usually have descriptive adjectives that may be followed by the names of the locations of the muscles. Examples: rectus (straight) femoris, rectus abdominus, and serratus (sawtooth) anterior.

✔ **Figurative names:** These muscle names are based on the muscles' resemblance to some objects. Examples: gastrocnemius (resembles the stomach) and trapezius (resembles a tablet).

Check out Table 6-1 for a rundown of prominent muscles in the body and key points to remember about each one.

Table 6-1	Muscles of the Body		
Muscle	*Origin*	*Insertion*	*Action*
Head			
Frontal	Galea aponeurotica	Eyebrow	Expression
Buccinator			Mastication
Orbicularis oculi	Encircles eye		Closes eye
Orbicularis oris	Encircles mouth		Closes mouth
Masseter	Zygoma	Mandible	Mastication
Temporalis	Temporal fossa	Mandible	Mastication
Zygomaticus	Zygoma	Corner of mouth	Smiling
Neck			
Sternocleidomastoid	Sternum, clavicle	Mastoid process of temporal bone	Rotation and flexion of the neck vertebrae
Back			
Latissimus dorsi	Vertebral column	Humerus	Extends at shoulder joint
Trapezius	Vertebral column	Clavicle, scapula	Rotates scapula
Pectoral girdle			
Pectoralis major	Sternum, clavicle	Humerus	Adduction shoulder joint
Shoulder			
Deltoid	Clavicle, scapula	Humerus	Abduction shoulder joint
Abdominal wall			
External abdominal oblique, internal abdominal oblique, transversus abdominus		Aponeurosis to linea alba	Stabilizes, protects, and supports internal viscera
Rectus abdominus	Pubis	Costal cartilage	Stabilizes, protects, and supports internal viscera
Thorax			
Diaphragm	Separates thoracic and abdominal cavities		Respiration
External intercostals	Between ribs		Respiration
Internal intercostals	Between ribs		Respiration

(continued)

Table 6-1 *(continued)*

Muscle	Origin	Insertion	Action
Arm			
Biceps brachii	Humerus, glenoid fossa of scapula	Radius	Flexion at elbow joint
Triceps brachii	Scapula, humerus	Olecranon of ulna	Extension of elbow joint
Flexor carpi radialis	Humerus	2nd to 3rd metacarpals	Flexor of wrist, abducts hand
Flexor carpi ulnaris	Humerus, ulna	5th metacarpal	Flexor of wrist, adducts hand
Supinator	Humerus, ulna	Radius	Supinates forearm
Extensor carpi ulnaris			
Extensor carpi radialis longus	Humerus	2nd metacarpal	Extends and abducts wrist
Extensor carpi radialis brevis	Humerus	3rd metacarpal	Extends and abducts wrist, steadies wrist during finger flexion
Leg			
Quadriceps			
Rectus femoris	Acetabulum	Tibia (patella)	Extends knee joint and flexes at hip
Vastus lateralis, vastus medialis, vastus intermedialis	Femur	Tibia	Extends knee joint and flexes at hip
Sartorius	Ilium	Tibia	Flexes at knee and hip
Adductors	Pubis	Femur	Adduction at hip joint
Gracilis	Pubis	Tibia	Adduction at hip joint
Hamstring group			
Biceps femoris	Ischium	Fibula	Flexion at knee joint
Semimembranosus	Ischium	Tibia	Flexion at knee joint
Semitendinosus	Ischium	Tibia	Flexion at knee joint
Gastrocnemius	Femur	Calcaneus by Achilles tendon	Flexion at knee and plantar
Soleus	Tibia	Calcaneus by Achilles tendon	Plantar flexion

Muscle	Origin	Insertion	Action
Hip			
Gluteus maximus	Ilium, sacrum, coccyx	Femur	Extends knee joint and flexes at hip

35. In the naming of the muscles, the latissimus dorsi, the rectus abdominis, and the serratus anterior are names based upon

    a. Shape

    b. Attachment

    c. Figurative name

    d. Function

36. In the naming of muscles, the sternocleidomastoid is based upon

    a. Function

    b. Location

    c. Attachment

    d. Figurative name

37. In humans, the origin of the biceps brachii would best include which of the following?

    a. Scapula

    b. Clavicle

    c. Fibula

    d. Ulna

38. Which of the following are insertions for the triceps and biceps brachii?

    a. Humerus and ulna

    b. Radius and humerus

    c. Scapula and humerus

    d. Radius and ulna

39.–43. Match the origins and insertions for the following muscles.

    39. _____ Semimembranosus   **a.** The pubis and the femur

    40. _____ Gracilis   **b.** The femur and the calcaneus

    41. _____ Sartorius   **c.** The ilium and the tibia

    42. _____ Gastrocnemius   **d.** The ischium and the tibia

    43. _____ Adductors   **e.** The pubis and the tibia

44.–48. Match the muscles with their actions.

    44. _____ Semitendinosus   **a.** Rotates scapula

    45. _____ Temporalis   **b.** Flexion of leg at knee joint

    46. _____ Biceps brachii   **c.** Extension at shoulder joint

    47. _____ Latissimus dorsi   **d.** Mastication

    48. _____ Trapezius   **e.** Flexion of arm

49.–53. Match the muscles with their locations.

49. _____ Latissimus dorsi	**a.** Head	
50. _____ Internal oblique	**b.** Abdomen	
51. _____ Quadriceps	**c.** Back	
52. _____ Masseter	**d.** Neck	
53. _____ Sternocleidomastoid	**e.** Thigh	

54. Which of the following is *not* included in the quadriceps group?

a. Vastus medialis

b. Vastus lateralis

c. Rectus abdominis

d. Rectus femoris

55. Where would you find the muscles called the biceps?

a. Arm

b. Neck

c. Leg

d. Back

e. Both a and c

56. What muscle divides the thoracic cavity from the abdominal cavity?

a. Diaphragm

b. External oblique

c. Transversus abdominis

d. Internal oblique

e. Rectus abdominis

57. The gastrocnemius and the soleus contribute to the

a. Dupuytren's contracture

b. Volkmann's contracture

c. Colles fracture

d. Klipped-Feil syndrome

e. Tendon of Achilles

58. Which of the following is *not* one of the muscles referred to as hamstrings?

a. Biceps femoris

b. Gracilis

c. Semimembranosus

d. Semitendinosus

Answers to Questions on Muscles

The following are answers to the practice questions presented in this chapter.

1   Which of the following is *not* a true statement? **a. Muscles represent 90 percent of the total body weight.** The average is less than half that, around 43 percent.

2   Muscle functions include **c. converting chemical energy into mechanical work.**

3   A necessary property for a muscle to perform work is **b. contractility.** If it doesn't contract, it's not a muscle.

4   The cellular unit in muscle tissue is the **c. fiber.** When it comes to muscle, fiber equals cell equals fiber.

5   A partial state of contraction, in part, defines **b. tonus.** That's the elusive muscle "tone" for the flabby amongst us.

6   It's possible to completely relax every muscle in the body. **b. False.** If every muscle in the body were to relax, the heart would stop beating and food would stop moving through the digestive system.

7   During embryonic development, tissue development is called **c. myogenesis.**

To remember this stage of development, combine the Greek *myo* with *genesis,* or new beginning.

8   Exercise forms new muscle fibers. **b. False.** Exercise can't form new fibers, but it can thicken what's already there.

9   This type of muscle tissue lacks cross-striations. **b. Smooth.** Without striations, this tissue can contract slowly and for a very long time.

10   Skeletal muscle fibers are encased in **a. a sarcolemma.** That's a thin membrane that helps move stimuli.

11   Which muscle type appears only in a single organ? **c. Cardiac.** And that sole organ is the heart.

12   Intercalated discs **b. may play a role in moving electrical impulses through the heart.** There's evidence that these structures help keep the heart synchronized.

13   Myosin: **d. A-band**

14   Segment of fibril from Z-line to Z-line: **e. Sarcomere**

15   Less-dense region of the A-band: **a. H-zone**

16   Structure to which filaments are attached: **b. Z-line**

17   Actin: **c. I-band**

18   Which of these terms doesn't belong in the following list? **e. Sarcolemma.** This is the membrane encasing the myofibrils. All the other answer options refer to various protein structures.

19   This part of a muscle doesn't change during contraction: **b. The A-bands.** All the other identified regions grow larger or smaller through the functions of myosin and actin.

20 A weak stimulus causes a muscle fiber to contract only partway. **b. False.** Muscle contractions are all-or-nothing; there's no such thing as a partial contraction.

21 Membrane covering a muscle fiber: **e. Sarcolemma**

22 Bundles of muscle fibers: **d. Fasciculi**

23 Connective tissue that surrounds a bundle of muscle fibers: **a. Perimysium**

24 Connective tissue through which arteries and veins enter muscle bundles: **c. Trabecula**

25 Flat, sheet-like tendon that serves as insertion for a large flat muscle: **b. Aponeurosis**

26 Muscles that tend to counteract or slow an action are called **a. antagonists.**

The muscles are against the action, so think of them as antagonistic.

27 Which of the following statements finishes this sentence and makes it *not* true: A contracting muscle unable to move a load **d. does no mechanical work and therefore doesn't develop any tension.** This statement is false because the contraction of the muscle causes tension in all cases.

28 A muscle contraction is referred to as **c. a twitch.** Keep in mind the twitch is the fundamental measure of muscle activity.

29 Which of the following in Figure 6-3 is a Class II lever? **b.**

30 Which of the following would provide the force in a Class III lever? **a. Biceps brachii.** It's the Popeye weight-lifting class, after all.

31 Which of the following would produce a wide range of movement with speed while sacrificing power? **b. Long weight arm, short power arm.** The longer the weight arm, the greater the range of action and speed *but* the less power there is.

32 Identify the bipennate bundle arrangement. **b. Rectus femoris.** Keep in mind the muscle naming conventions.

33 Which of the following is considered a dissectable connective tissue? **c. Fasciae.** This connective tissue can be found in most of the body.

34 The most extensive fascia in the body is **b. deep.**

35 In the naming of the muscles, the latissimus dorsi, the rectus abdominis, and the serratus anterior are names based upon **a. shape.** *Latissimus* stems from the Latin for "wide," *rectus* from the Latin word for "straight," and *serratus* from the Latin word for "notched" or "scalloped."

36 In the naming of muscles, the sternocleidomastoid is based upon **c. attachment.** You can figure out this answer by recalling that *cleido* stems from the Latin word for "collarbone."

37 In humans, the origin of the biceps brachii would best include which of the following? **a. Scapula**

38 Which of the following are insertions for the triceps and biceps brachii? **d. Radius and ulna**

39   Semimembranosus: **d. The ischium and the tibia**

40   Gracilis: **e. The pubis and the tibia**

41   Sartorius: **c. The ilium and the tibia**

42   Gastrocnemius: **b. The femur and the calcaneus**

43   Adductors: **a. The pubis and the femur**

44   Semitendinosus: **b. Flexion of leg at knee joint**

45   Temporalis: **d. Mastication**

46   Biceps brachii: **e. Flexion of arm**

47   Latissimus dorsi: **c. Extension at shoulder joint**

48   Trapezius: **a. Rotates scapula**

49   Latissimus dorsi: **c. Back**

50   Internal oblique: **b. Abdomen**

51   Quadriceps: **e. Thigh**

52   Masseter: **a. Head**

53   Sternocleidomastoid: **d. Neck**

54   Which of the following is *not* included in the quadriceps group? **c. Rectus abdominis**. The term *abdominis* is the giveaway here because it should make you think of the abdomen. The rest of the quadriceps group is found in the upper leg, where they belong.

55   Where would you find the muscles called the biceps? **e. Both a and c (arm and leg)**. Biceps is a muscle with two heads or points of origin. Although people usually think of the biceps brachii in the arm, you can't forget about the biceps femoris at the back of the thigh.

56   What muscle divides the thoracic cavity from the abdominal cavity? **a. Diaphragm.** And without it, you couldn't breathe.

57   The gastrocnemius and the soleus contribute to the **e. tendon of Achilles.** The soleus lies under the gastrocnemius in the calf of each leg.

58   Which of the following is *not* one of the muscles referred to as hamstrings? **b. Gracilis.** The other three answer options all are listed as hamstring muscles.

Chapter 7

It's Skin Deep: The Integumentary System

Did you know that the skin is the body's largest organ? In an average person, its 17 to 20 square feet of surface area represents 15 percent of the body's weight. Self-repairing and surprisingly durable, the skin is the first line of defense against the harmful effects of the outside world. It helps retain moisture; regulates body temperature; hosts the sense receptors for touch, pain, and heat; excretes excess salts and small amounts of waste; and even stores blood to be moved quickly to other parts of the body when needed.

Skin is jam-packed with components; it has been estimated that every square inch of skin contains 15 feet of blood vessels, 4 yards of nerves, 650 sweat glands, 100 oil glands, 1,500 sensory receptors, and over 3 million cells with an average lifespan of 26 days that are constantly being replaced.

In this chapter, we peel back the surface of this most-visible organ system. We also give you plenty of opportunities to test your knowledge.

Dermatology Down Deep

Skin — together with hair, nails, and glands — composes the *integumentary system* (shown in Figure 7-1). The name stems from the Latin verb *integere*, which means "to cover." The relevant Greek and Latin roots include *dermato* and *cutis*, both of which mean "skin."

The skin consists of two primary parts: the *epidermis* and the *dermis*. (Recall the Greek root *epi–*, which means "upon" or "above.") Underlying the epidermis and dermis is the *hypodermis* or *superficial fascia* (also sometimes called *subcutaneous tissue*), which acts as a foundation but is not part of the skin. Composed of areolar (porous) and adipose (fat) tissue, it anchors the skin through fibers that extend from the dermis. Underneath, the hypodermis attaches loosely to tissues and organs so that muscles can move freely. Around elbow and knee joints, the hypodermis contains fluid-filled sacs called *bursae*. The fat in the hypodermis buffers deeper tissues and acts as insulation, preventing heat loss from within the body's core. The hypodermis also is home to pressure-sensitive nerve endings called *lamellated* or *Pacinian corpuscles* that respond to a deeper poke in the skin.

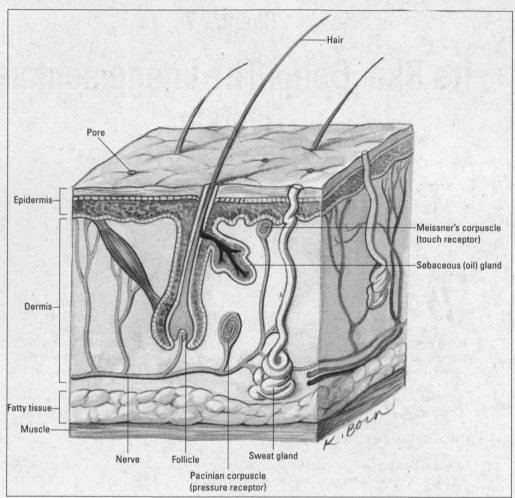

Figure 7-1:
The integu-
mentary
system.

Epidermis: Don't judge this book by its cover

Epidermis, which contains no blood vessels, is made up of layers of closely packed epithelial cells. From the outside in, these layers are the following:

- ✔ **Stratum corneum** (literally the "horny layer") is about 20 layers of flat, scaly, dead cells containing a type of water-repellent protein called *keratin*. These cells, which represent about three-quarters of the thickness of the epidermis, are said to be *cornified,* which means that they're tough and horny like the cells that form hair or fingernails. Humans shed this layer of tough, durable skin at a prodigious rate; in fact, much of household dust consists of these flaked-off cells. Where the skin is rubbed or pressed more often, cell division increases, resulting in calluses and corns.

- ✔ **Stratum lucidum** (from the Latin word for "clear") is found only in the thick skin on the palms of the hands and the soles of the feet. This translucent layer of dead cells contains *eleidin,* a protein that becomes keratin as the cells migrate into the stratum corneum, and it consists of cells that have lost their nuclei and cytoplasm.

- ✔ **Stratum granulosum** is three to five layers of flattened cells containing *keratohyalin,* a substance that marks the beginning of keratin formation. No nourishment from blood vessels reaches this far into the epidermis, so cells are either dead or dying by the time they reach the stratum granulosum. The nuclei of cells found in this layer are degenerating; when the nuclei break down entirely, the cell can't metabolize nutrients and dies.

- ✔ **Stratum spinosum** (also sometimes called the *spinous layer*) has ten layers containing *prickle cells,* named for the spine-like projections that connect them with other cells in the layer. *Langerhans cells,* believed to be involved in the body's immune response, are prevalent in the upper portions of this layer and sometimes the lower part of the stratum granulosum; they migrate from the skin to the lymph nodes in response to infection. Some mitosis (cell division) takes place in the stratum spinosum, but the cells lose the ability to divide as they mature.

- ✔ **Stratum basale** (or *stratum germinativum*) is also referred to as the *germinal layer* because this single layer of mostly columnar stem cells generates all the cells found in the other epidermal layers. It rests on the papillary (rough or bumpy) surface of the dermis, close to the blood supply needed for nourishment and oxygen. The mitosis that constantly occurs here replenishes the skin; it takes about two weeks for the cells that originate here to migrate up to the stratum corneum, and it's another two weeks before they're shed. About a quarter of this layer's cells are *melanocytes,* cells that synthesize a pale yellow to black pigment called *melanin* that contributes to skin color and provides protection against ultraviolet radiation (the kind of radiation found in sunlight). The remaining cells in this layer become *keratinocytes,* the primary epithelial cell of the skin. Melanocytes secrete melanin directly into the keratinocytes in a process called *cytocrine secretion.* Merkel's cells, a large oval cell believed to be involved in the sense of touch, occasionally appear amid the keratinocytes.

In addition to melanin, the epidermis contains a yellowish pigment called *carotene* (the same one found in carrots and sweet potatoes). Found in the stratum corneum and the fatty layers beneath the skin, it produces the yellowish hue associated with Asian ancestry or increased carrot consumption. The pink to red color of Caucasian skin is caused by *hemoglobin,* the red pigment of the blood cells. Because Caucasian skin contains relatively less melanin, hemoglobin can be seen more easily through the epidermis. Sometimes the limited melanin in Caucasian skin pools in small patches. Can you guess the name of those patches of color? Yep, they're freckles. Albinos, on the other hand, have no melanin in their skin at all, making them particularly sensitive to ultraviolet radiation.

Ridges and grooves form on the outer surface of the epidermis to increase the friction needed to grasp objects or move across slick surfaces. On hands and feet, these ridges form patterns of loops and whorls — fingerprints, palm prints, and footprints — that are unique to each person. You leave these imprints on smooth surfaces because of the oily secretions of the sweat glands on the skin's surface. In addition to these finer patterns, the areas around joints develop patterns called *flexion lines.* Deeper and more permanent lines are called *flexion creases.*

Dermis: It's more than skin deep

Beneath the epidermis is a thicker, fibrous structure called the dermis, or *corium.* It consists of the following two layers, which blend together:

- ✔ The outer, soft *papillary layer* contains elastic and reticular (net-like) fibers that project into the epidermis to bring blood and nerve endings closer. *Papillae* (finger-like projections) containing loops of capillaries increase the surface area of the dermis and anchor the epidermis. Some of these papillae contain

> *Meissner's corpuscles,* nerve endings that are sensitive to soft touch. It's the dermal papillae that form the epidermal ridges referred to as fingerprints.
>
> ✔ The inner, thicker *reticular layer* (from the Latin word *rete* for "net") is made up of dense, irregular connective tissue containing interlacing bundles of collagenous and elastic fibers that form the strong, resistant layer used to make leather and suede from animal hides. This layer is what gives skin its strength, extensibility, and elasticity. Within the reticular layer are *sebaceous glands* (oil glands), sweat glands, fat cells, and larger blood vessels.

Cells in the dermis include *fibroblasts* (from which connective tissue develops), *macrophages* (which engulf waste and foreign microorganisms), and adipose tissue. Thickest on the palms of the hands and soles of the feet, the dermis is thinnest over the eyelids, penis, and scrotum. It's thicker on the back (posterior) than on the stomach (anterior) and thicker on the sides (lateral) than toward the middle of the body (medial). The various skin "accessories" — blood vessels, nerves, glands, and hair follicles — are embedded here.

See if you've got the skinny on skin so far:

Q. The layer of epidermis that's composed of a horny, cornified tissue that sloughs off is called the stratum

   a. Corneum

   b. Lucidum

   c. Granulosum

   d. Spinosum

   e. Basale

A. The correct answer is corneum. Cornified, corns — think of how hard a kernel of popcorn can be.

1. The layer of epidermis in which mitosis takes place is the stratum

   a. Corneum

   b. Lucidum

   c. Granulosum

   d. Papilla

   e. Basale

2. The papillary layer of the dermis

   a. Is composed of numerous projections

   b. Extends into the epidermis

   c. Carries the blood and nerve endings close to the epidermis

   d. Aids in holding the epidermis and dermis together

   e. All of the above

3. The function of the epidermal ridges on the fingers is to

    a. Provide a means of identification

    b. Increase the friction of the epidermal surface

    c. Decrease water loss by the tissues

    d. Aid in regulating body temperature

    e. Prevent bacterial infection

4. The color of Caucasian skin is due to

    a. Carotene pigment in the dermis

    b. The high level of melanin in the epidermis

    c. Less melanin in the skin, allowing the blood pigment to be seen

    d. The absence of all pigment

    e. Melanin and carotene pigments

5. The sequence of layers in the epidermis from the dermis outward is

    a. Corneum, lucidum, granulosum, spinosum, basale

    b. Corneum, granulosum, lucidum, basale, spinosum

    c. Spinosum, basale, granulosum, corneum, lucidum

    d. Basale, spinosum, granulosum, lucidum, corneum

    e. Basale, lucidum, corneum, spinosum, granulosum

6. The subcutaneous layer of tissue can be called the

    a. Epidermis

    b. Superficial fascia

    c. Papillary layer

    d. Inner reticular layer

    e. Dermis

7. The lines in fingerprints are determined by

    a. The contours of the dermal papillae

    b. The thickness of the dermis

    c. The bundles of collagenous and elastic fibers in the epidermis

    d. The fibroblasts, macrophages, and adipose tissue surrounding the nerves

    e. The surface layer of cells that are constantly being shed

8. A function *not* pertaining to the skin is

    a. Aids in retaining water

    b. Regulates body temperature

    c. Contains sense receptors

    d. Excretes some waste materials

    e. Provides movement

9. A layer of dense, irregular connective tissue containing interlacing bundles of collagenous and elastic fibers is the

   a. Basale layer of the epidermis

   b. Reticular layer of the dermis

   c. Outer layer of the hypodermis

   d. Papillary layer of the dermis

   e. Inner layer of the hypodermis

10. The stratum corneum cells contain a tough, water-repellant protein called

    a. Keratohyalin

    b. Eleidin

    c. Keratin

    d. Cerumen

    e. Sweat

11. The epidermal layer containing keratohyalin is the

    a. Stratum germinativum

    b. Stratum spinosum

    c. Stratum lucidum

    d. Stratum granulosum

    e. Stratum corneum

12. The layer of skin attached to the superficial fascia is the

    a. Papillary layer of the dermis

    b. Stratum granulosum

    c. Stratum germinativum

    d. Holorine layer of the epidermis

    e. Reticular layer of the dermis

13. Flattened and irregular cells with small, spine-like projections that connect them with other cells in the layer are referred to as

    a. Prickle cells

    b. Langerhans cells

    c. Melanocytes

    d. Merkel's cells

    e. Keratohyalin cells

14. If melanin forms into patches, it's referred to as

    a. Flexion creases

    b. Freckles

    c. Dermal papillae

    d. Lamellated corpuscles

    e. Matrix

Touching a Nerve in the Integumentary System

At least four kinds of receptors are involved in creating the sensation of touch.

- ✔ **Free nerve endings:** These afferent nerve endings are *dendrites* (branched extensions) of sensory neurons that act primarily as pain receptors, although some sense temperature, touch, and muscles (including the sensation of "stretch"). Found all over the body, free nerve endings are especially prevalent in epithelial and connective tissue. These small-diameter fibers have a swelling at the end that responds to touch and sometimes heat, cold, or pain. Some of the endings are disc-shaped structures called *Merkel discs* that function as light-touch receptors within the deep layers of the epidermis.

- ✔ **Meissner's corpuscles:** These light-touch mechanoreceptors lie within the dermal papillae. They're small, egg-shaped capsules of connective tissue surrounding a spiraled end of a dendrite. Most abundant in sensitive skin areas such as the lips and fingertips, these corpuscles and free nerve endings can sense a quick touch but not a sustained one. That's why your skin is able to ignore the touch sensation of your own clothing.

- ✔ **Pacinian corpuscles:** These deep-pressure mechanoreceptors are dendrites surrounded by concentric layers of connective tissue. Found deep within the dermis, they respond to deep or firm pressure and vibrations. Each is over 2 millimeters long and therefore visible to the naked eye.

- ✔ **Hair nerve endings:** These mechanoreceptors respond to a change in position of a hair. They consist of bare dendrites.

There are two primary temperature receptors, one for heat and one for cold:

- ✔ **End-bulbs of Krause:** Also known as *Krause's corpuscles,* these cold receptors usually activate below 68 degrees F (20 degrees C). They consist of a bulbous capsule surrounding the dendrite and are commonly found throughout the body in the dermis as well as in the lips, the tongue, and the conjunctiva of the eyes.

- ✔ **Brushes of Ruffini:** Also known as *Ruffini cylinders* or *Ruffini's corpuscles,* these warmth receptors respond to temperatures between 77 degrees and 113 degrees F (25 degrees to 45 degrees C). Found primarily in the dermis and subcutaneous tissue, they're dendrite endings enclosed in a flattened capsule. Because there are fewer of them than Krause's end-bulbs and because they lie in deeper tissue, the body is less sensitive to heat than to cold.

In addition to the receptors for touch and temperature, the dermis has neuromuscular spindles (also called *proprioceptors*) that transmit information to the spinal cord and brain about the lengths and tensions of muscles. This information helps provide awareness of the body's position and the relative position of body parts. The spindles also assist with muscle coordination and muscle action efficiency.

Test whether you're staying in touch with this section:

15. The sensation of a soft touch is received by

   a. Ruffini's corpuscles

   b. Pacinian corpuscles

   c. The Crypt of Lieberkuhn

   d. The end-bulbs of Krause

   e. Meissner's corpuscles

16. End-bulbs of Krause are receptors for cold that usually are activated at temperatures below

   a. 98.6 degrees F

   b. 20 degrees F

   c. 68 degrees F

   d. 45 degrees F

   e. 77 degrees F

Accessorizing with Hair, Nails, and Glands

Mother Nature has accessorized your fashionable over-wrap with a variety of specialized structures that grow from the epidermis: hair, fingernails, toenails, sebaceous glands, and sweat glands.

Wigging out about hair

Like most mammals, hair covers the entire human body except for the lips, eyelids, palms of the hands, soles of the feet, nipples, and portions of external reproductive organs. But human body hair generally is sparser and much lighter in color than that sported by most other mammals. Animals have hair for protection and temperature control. For humans, however, body hair is largely a secondary sex characteristic. A thick head of hair protects the scalp from exposure to the sun's harmful rays and limits heat loss. Eyelashes block sunlight and deflect debris from the eyes. Hair in the nose and ears prevents airborne particles and insects from entering. Touch receptors connected to hair follicles respond to the lightest brush. The average adult has about 5 million hairs, with about 100,000 of those growing from the scalp. Normal hair loss from an adult scalp is about 70 to 100 hairs each day, although baldness can result from genetic factors, hormonal imbalances, scalp injuries, disease, dietary deficiencies, radiation, or chemotherapy.

Each hair grows at an angle from a follicle embedded in the epidermis and extending into the dermis; scalp hairs sometimes reach as far as the hypodermis. Nerves reach the hair at the follicle's expanded base, called the *bulb,* where a nipple-shaped papilla of connective tissue and capillaries provide nutrients to the growing hair. Epithelial cells in the bulb divide to produce the hair's *shaft* (the part that extends out of the follicle). (The part of the hair within the follicle is called the *root.*) The shape of a hair's cross section can vary from round to oval or even flat; oval hairs grow out appearing wavy or curly, flat hairs appear kinky, and round hairs grow out straight. Each scalp hair grows for two to three years at a rate of about ⅓ to ½ millimeter per day, or 10 to 18 centimeters per year. When mature, the hair rests for three or four months before slowly losing its attachment. Eventually, it falls out and is replaced by a new hair.

Hair pigment (which is melanin, just as in the skin) is produced by melanocytes in the follicle and transferred to the hair's cortex and medulla cells. Three types of melanin — black, brown, and yellow — combine in different quantities for each individual to produce different hair colors ranging from light blonde to black. Gray and white hairs grow in when melanin levels decrease and air pockets form where the pigment used to be.

Wondering why you have to shampoo so often? Hair becomes oily over time thanks to *sebum,* a mixture of cholesterol, fats, and other substances secreted from a *sebaceous* (or *holocrine*) gland found next to each follicle. Sebum keeps both hair and skin soft, pliable, and waterproof. Attached to each follicle is a smooth muscle called an *arrector pili* (literally "raised hair") that both applies pressure to the sebaceous gland and straightens the hair shaft, depressing the skin in a pattern called *goose bumps* or *goose pimples*.

Each hair is made up of three concentric layers of keratinized cells:

- A central core, called the *medulla,* consists of large cells containing eleidin that are separated by air spaces; in fine hair, the medulla may be small or entirely absent.

- A *cortex* surrounding the medulla forms the major part of the hair shaft with several layers of flattened cells. The cortex also has elongated pigment-bearing cells in dark hair and air pockets in white hair.

- The outermost *cuticle* is a single layer of overlapping cells with the free end pointing upward. The cuticle strengthens and compacts the inner layers, but abrasion tends to wear away the end of the shaft, exposing the medulla and cortex in a pattern known as *split ends.*

Nailing the fingers and toes

Human nails (which actually are vestigial claws) have three parts: a *root bed* at the nail base, a *body* that's attached to the fingertip, and a *free edge* that grows beyond the end of the finger or toe. Heavily cornified tissue forms the nails from modified strata corneum and lucidum. A narrow fold of the stratum corneum turns back to form the *eponychium,* or *cuticle.* Under the nail, the nail bed is formed by the strata basale and spinosum. At the base of the nail, partially tucked under the cuticle, the strata thicken to form a whitish area called the *lunula* (literally "little moon") that can be seen through the nail. Beneath the lunula is the *nail matrix,* a region of thickened strata where mitosis pushes previously formed cornified cells forward, making the nail grow. Under the free edge of the nail, the stratum corneum thickens to form the *hyponychium.* Nails are pinkish in color because of hemoglobin in the underlying capillaries, which are visible through the translucent cells of the nail.

On average, fingernails grow about 1 millimeter each week. Toenails tend to grow even more slowly. Nails function as an aid to grasping, as a tool for manipulating small objects, and as protection against trauma to the ends of fingers and toes.

Sweating the details

Humans perspire over nearly every inch of skin, but anyone with sweaty palms or smelly feet can attest to the fact that sweat glands are most numerous in the palms and soles, with the forehead running a close third. There are two types of sweat, or *sudoriferous,* glands: *eccrine* and *apocrine.* Both are coiled tubules embedded in the dermis or subcutaneous layer composed of simple columnar cells.

Eccrine glands are distributed widely over the body — an average adult has roughly 3 million of them — and produce the watery, salty secretion you know as sweat. Each gland's duct passes through the epidermis to the skin's surface, where it opens as a *sweat pore.* The sympathetic division of the autonomic nervous system controls when and how much perspiration is secreted depending on how hot the body becomes. Sweat helps cool the skin's surface by evaporating as fast as it forms. About 99 percent of eccrine-type sweat is water, but the remaining 1 percent is a mixture of sodium chloride and other salts, uric acid, urea, amino acids, ammonia, sugar, lactic acid, and ascorbic acid.

Apocrine sweat glands are located primarily in armpits (known as the *axillary region*) and the groin area. Usually associated with hair follicles, they produce a white, cloudy secretion that contains organic matter. Although apocrine-type sweat contains the same basic components as eccrine sweat and also is odorless when first secreted, bacteria quickly begin to break down its additional fatty acids and proteins — explaining the post-exercise underarm stench. In addition to exercise, sexual and other emotional stimuli can cause contraction of cells around these glands, releasing sweat.

Getting an earful

The occasionally troublesome yellowish substance known as earwax is secreted in the outer part of the ear canal from modified sudoriferous glands called *ceruminous glands* (the Latin word *cera* means "wax"). Lying within the subcutaneous layer of the ear canal, these glands have ducts that either open directly into the ear canal or empty into the ducts of nearby sebaceous glands. Technically called *cerumen,* earwax is the combined secretion of these two glands. Working with ear hairs, cerumen traps any foreign particles before they reach the eardrum. As the cerumen dries, it flakes and falls from the ear, carrying particles out of the ear canal.

Think you've got a grip on everything to do with hair, nails, and glands? Find out by answering the following practice questions:

17. The cuticle is also called the

   a. Lunula

   b. Hyponychium

   c. Eponychium

   d. Nail matrix

   e. Perinychium

18. Perspiration is formed by the

   a. Sebaceous glands

   b. Ceruminous glands

   c. Endocrine glands

   d. Merkel's glands

   e. Sudoriferous glands

19. The cause of graying hair is

   a. Production of melanin in the shaft of the hair

   b. Production of carotene in the shaft of the hair

   c. Decrease in blood supply to the hair

   d. Lack of melanin in the shaft of the hair

   e. Parenthood

20. Hair develops from a(n)

    a. Arrector pili

    b. Shaft

    c. Follicle

    d. Sebaceous gland

    e. Lanugo

21. The nails are modifications of the epidermal layers

    a. Corneum and lucidum

    b. Lucidum and granulosum

    c. Granulosum and spinosum

    d. Spinosum and basale

22. The muscle that straightens a hair and puts pressure on a gland causing it to secrete is the

    a. Terminalis muscle

    b. Arrector pili muscle

    c. Internal oblique muscle

    d. External transversus muscle

    e. Internal rectus muscle

23. A factor not associated with baldness is

    a. Genetics

    b. Hormonal imbalances

    c. Scalp injuries

    d. Lack of carotene

    e. Disease

24. Sebaceous glands

    a. Produce a watery solution called sweat

    b. Produce an oily mixture of cholesterol, fats, and other substances

    c. Produce a waxy secretion called cerumen

    d. Accelerate aging

    e. Are associated with endocrine glands

25. The bulb of the follicle of a hair contains epithelial cells (germinating cells) that are continuous with the

    a. Papillary layer of the dermis

    b. Stratum corneum

    c. Stratum germinativum

    d. Reticular layer of the dermis

    e. Statum lucidum

26. Cooling of the skin's surface is aided by the

   a. Endocrine glands

   b. Sebaceous glands

   c. Ceruminous glands

   d. Prickle glands

   e. Sweat glands

27. This gland contains true sweat, fatty acids, and proteins, and acquires an unpleasant odor when bacteria breaks down the organic molecules it secretes.

   a. Apocrine sweat gland

   b. Sebaceous gland

   c. Ceruminous gland

   d. Eccrine sweat gland

   e. Mammary gland

28. Which of the following is true about fingernails?

   a. They're derived from the hypodermis.

   b. They contain carotene.

   c. They grow more slowly than toenails.

   d. They grow about 1 millimeter per week.

   e. They're not a skin accessory.

29. The gland that secretes an oily mixture of cholesterol, fats, and other substances into hair follicles to keep hair and skin soft, pliable, and waterproof is the

   a. Sweat gland

   b. Sebaceous gland

   c. Ceruminous gland

   d. Apocrine gland

   e. Eccrine gland

Answers to Questions on the Skin

The following are answers to the practice questions presented in this chapter.

1. The layer of epidermis in which mitosis takes place is the stratum **e. basale.**

This layer also is called the stratum germinativum, but a simpler memory tool is simply to associate it with the "base" of the epidermis.

2. The papillary layer of the dermis **e. all of the above.** Busy little finger-like projections, those papillae.

3. The function of the epidermal ridges on the fingers is to **b. increase the friction of the epidermal surface.** It's Mother Nature's way of helping you cling to tree branches or grab food.

4. The color of Caucasian skin is due to **c. less melanin in the skin, allowing the blood pigment to be seen.** Here's a fun experiment: Turn off the lights, press your fingers together, and hold a flashlight under them. See the red glow? That's hemoglobin, too.

5. The sequence of layers in the epidermis from the dermis outward is **d. basale, spinosum, granulosum, lucidum, corneum.**

Memory tool time: Base, spine, grain, Lucy, corny. Or try the first letters of Be Super Greedy, Less Caring. Insensitive, yes, but effective.

6. The subcutaneous layer of tissue can be called the **b. superficial fascia.** Subcutaneous is the same as hypodermis (from the Greek *hypo*– for "beneath").

7. The lines in fingerprints are determined by **a. the contours of the dermal papillae.**

8. A function *not* pertaining to the skin is **e. provides movement.** That would be in the realm of muscles.

9. A layer of dense, irregular connective tissue containing interlacing bundles of collagenous and elastic fibers is the **b. reticular layer of the dermis.** The description in this question sounds like a tough structure, so it may help you to remember that the reticular layer is what's used to make leather from animal hides.

10. The stratum corneum cells contain a tough, water-repellant protein called **c. keratin.** Associate the words "corneum" and "keratin," and you're in great shape.

11. The epidermal layer containing keratohyalin is the **d. stratum granulosum.** Keratohyalin eventually becomes keratin, so think of the layer where the cells are starting to die off.

12. The layer of skin attached to the superficial fascia is the **e. reticular layer of the dermis.** Reticular means net-like; it makes sense that this netting lies between the dermis and the hypodermis.

13. Flattened and irregular cells with small, spine-like projections that connect them with other cells in the layer are referred to as **a. prickle cells.** The spines make them look prickly, hence the name.

14. If melanin forms into patches, it's referred to as **b. freckles.** Ever noticed how kids have more freckles at the end of a long summer spent outdoors? That's ultraviolet radiation working on those melanin patches.

15 The sensation of a soft touch is received by **e. Meissner's corpuscles.** While it's true that several different nerves are involved in the overall sense of touch, the Meissner's are the most responsive to touch.

16 End-bulbs of Krause are receptors for cold that usually are activated at temperatures below **c. 68 degrees F.**

TIP Specific temperatures may seem tough to remember, but look at it this way: When it's 45 degrees F, you definitely need a jacket. When it's 77 degrees F, you don't. But when it's 68 degrees F, you'll want to carry a light jacket in case it gets colder. Voila! A cold receptor activation temperature!

17 The cuticle is also called the **c. eponychium.** Recall that the prefix *ep–* refers to "upon" or "around," whereas the prefix *hypo–* refers to "below" or "under." The cuticle is around the base of the nail, so it's the eponychium, not the hyponychium.

18 Perspiration is formed by the **e. sudoriferous glands.** The Latin word *sudor* means "sweat."

19 The cause of graying hair is **d. lack of melanin in the shaft of the hair.** Despite the medical cause, people often suspect that answer "e. parenthood" has a lot to do with graying hair.

20 Hair develops from a(n) **c. follicle.** The Latin translation of this word is "small cavity" or "sac," so it makes sense that this would be an origination place.

21 The nails are modifications of the epidermal layers **a. corneum and lucidum.** These are the two upper layers.

22 The muscle that straightens a hair and puts pressure on a gland causing it to secrete is the **b. arrector pili muscle.** "Arrector" is similar to "erector" (in fact, the muscle is sometimes called that), which implies straightening, and the Latin word for "hair" is *pili*.

23 A factor not associated with baldness is **d. lack of carotene.** This answer just means that your hair won't turn orange, not necessarily that it will fall out of your scalp.

24 Sebaceous glands **b. produce an oily mixture of cholesterol, fats, and other substances.** That secretion's called sebum — hence "sebaceous glands."

25 The bulb of the follicle of a hair contains epithelial cells (germinating cells) that are continuous with the **c. stratum germinativum.** Germinating cells from the germinativum. Don't forget, though, that this layer also is called the stratum basale, or base stratum.

26 Cooling of the skin's surface is aided by the **e. sweat glands.** The evaporation of perspiration cools the skin naturally.

27 This gland contains true sweat, fatty acids, and proteins, and acquires an unpleasant odor when bacteria breaks down the organic molecules it secretes. **a. Apocrine sweat gland.** These are the truly stinky sweat glands.

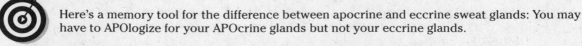

TIP Here's a memory tool for the difference between apocrine and eccrine sweat glands: You may have to APOlogize for your APOcrine glands but not your eccrine glands.

28 Which of the following is true about fingernails? **d. They grow about 1 millimeter per week.** The other answer options are either just plain wrong or nonsensical.

29 The gland that secretes an oily mixture of cholesterol, fats, and other substances into hair follicles to keep hair and skin soft, pliable, and waterproof is the **b. sebaceous gland.** Sebum, sebaceous, oily — just don't call it sweat.

Part III
Feed and Fuel: Supply and Transport

The 5th Wave   By Rich Tennant

"It's important that you get your high blood pressure under control. It could not only affect your kidney and heart, but also your current brain."

In this part . . .

There are a few things humans just can't live without.
This part gives you a rundown of how your body gets
what it needs and what it does to take advantage of these
resources.

Each of the chapters in this part delves into a different
major body system, starting with the respiratory system
and what a few deep breaths can do for the human
machine. Next up is the digestive system, fueling the
system with food; you follow a mouthful of food from its
entry in the mouth to expulsion of waste after every possible nutrient has been wrung from it. We check in on the circulatory system and its blood-filled internal transit routes
that carry both nutrients and oxygen to every nook and
cranny of the body. Then it's on to the lymphatic system's
distribution of crucial immune system functions. Of course,
all this supply and transport is bound to lead to a waste
issue; we close out this part with a look at how the urinary
system collects the body's trash and dispenses with it.

Chapter 8

Oxygenating the Machine: The Respiratory System

*P*eople need lots of things to survive, but the most urgent need from moment to moment is oxygen. Without a continual supply of this vital element, we don't last long. But if we have reserves of the other things we need — carbohydrates, fats, and proteins — why don't we have some kind of storehouse of oxygen, too? Simple. It's readily available in the air around us, so we've never needed to evolve a means for storing it. Nonetheless, our stored food supplies would be useless without oxygen; our bodies can't metabolize the energy they need from these substances without a constant stream of oxygen to keep things percolating along. All that metabolizing creates another equally important need, however. We must have a means for getting rid of our bodies' key gaseous waste — carbon dioxide, or CO_2. If it builds up in our systems, we die. It must be removed from our bodies almost as fast as it's formed. Conveniently, breathing in fulfills our need for oxygen and breathing out fulfills our need to expel carbon dioxide.

In this chapter, you get a quick review of Mother Nature's dual-purpose system and plenty of opportunities to test your knowledge about the lungs and other parts of the respiratory system.

Breathing In Oxygen, Breathing Out CO_2

Respiration, or the exchange of gases between an organism and its environment, occurs in three distinct processes:

- **Breathing:** The technical term is *pulmonary ventilation,* or the movement of air into and out of the lungs. (Breathing is also called *inspiration* and *expiration.*)

- **Exchanging gases:** This takes place between the alveolar cells in the lungs, the blood, and the body's cells in two ways:

  - **Pulmonary, or external, respiration:** The exchange in the lungs when blood gains oxygen and loses carbon dioxide, transforming it from venous blood into arterial blood

  - **Systemic, or internal, respiration:** The exchange within systemic capillaries when the blood releases some of its oxygen and collects carbon dioxide from the tissues

- **Cellular respiration:** Oxygen is used in the catabolism of substances like glucose for the production of energy (see Chapter 1).

Here are some key respiratory terms to keep in mind:

- **Adult breathing rate:** About 12 to 20 times per minute.

- **Anoxia:** Oxygen deficiency in which the cells either don't have or can't utilize sufficient oxygen to perform normal functions.

- **Asphyxia:** Lack of oxygen with an increase in carbon dioxide in the blood and tissues; accompanied by a feeling of suffocation leading to coma.

- **Expiration** or **exhalation:** The diaphragm returns to its domed shape as the muscle fibers relax, via elastic recoil of the lungs and tissues lining the thoracic cavity, the external intercostal muscles relax, and the internal intercostal muscles contract. This movement pulls the ribs back into place, decreasing the volume of the thoracic cavity and increasing pressure, forcing air out of the lungs.

- **Hypoxia:** Low oxygen content in the inspired air.

- **Inspiration** or **inhalation:** When the muscles of the diaphragm contract, its dome shape flattens; simultaneously, the contraction of the external intercostal muscles pulls the ribs upward and increases the volume of the thoracic cavity, decreasing the intra-alveolar pressure. The pressure difference between the atmosphere and the lungs diffuses air into the respiratory tract.

- **Lung capacity:** The vital capacity plus the residual air.

- **Mediastinum:** The region between the lungs extending from the sternum ventrally (at the front) to the thoracic vertebrae dorsally (at the back), and superiorly (top) from the entrance of the thoracic cavity to the diaphragm inferiorly (at the bottom).

- **Minimal air:** The volume of air in the lungs when they're completely collapsed (150 cubic centimeters in an adult).

- **Phrenic nerve:** The nerve that innervates (stimulates) the diaphragm.

- **Residual air:** The volume of air remaining in the lungs after the most forceful expiration (1,200 cubic centimeters in an adult).

- **Respiratory centers:** Nerve centers for regulating breathing located in the *medulla oblongata,* or brain stem. The centers are influenced by the amount of carbon dioxide in the blood.

- **Tidal air:** The volume of air inspired and expired in the resting state (500 cubic centimeters in an adult).

- **Vital capacity:** The volume of air moved by the most forceful expiration after a maximum inspiration. It represents the total moveable air in the lungs (4,600 cubic centimeters in an adult).

Here's what happens as you breathe in and out (see Figure 8-1): Red blood cells use a pigment called *hemoglobin* to carry oxygen and carbon dioxide throughout the body through the circulatory system (for more on that system, turn to Chapter 10). Hemoglobin bonds loosely with oxygen, or O_2, to carry it throughout the body; the bonded hemoglobin is called *oxyhemoglobin.*

After hemoglobin releases its oxygen molecules, it picks up carbon dioxide, or CO_2, to deliver to the lungs for exhalation. The freshly bonded hemoglobin becomes *carbohemoglobin* (*carbhemoglobin* or *carbaminohemoglobin*).

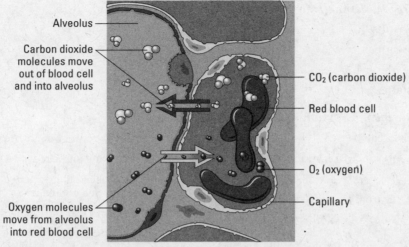

Alveolus

Carbon dioxide molecules move out of blood cell and into alveolus

CO_2 (carbon dioxide)

Red blood cell

O_2 (oxygen)

Capillary

Oxygen molecules move from alveolus into red blood cell

Figure 8-1: Oxygen-carbon dioxide exchange in lung.

Illustration by LifeART Image Copyright © 2007. Wolters Kluwer Health - Lippincott Williams & Wilkins.

See whether you're carrying away enough information about respiration by tackling the following practice questions:

Q. The air that moves in and out of the lungs during normal, quiet breathing is called

   a. Tidal volume, or tidal air

   b. Inspiratory reserve

   c. Vital capacity

   d. Lung capacity

A. The correct answer is tidal volume, or tidal air. The question asks only about air moved during normal, quiet breathing, not the kind of forceful air movement involved in measuring lung capacity. Think of the normal ebb and flow of the ocean's tide as opposed to the waves of a raging storm.

1. Which of the following gases are dissolved and held in chemical combination in the blood?

   a. Nitrogen and CO_2

   b. Chlorine and CO

   c. Oxygen and CO_2

   d. Nitrogen and O_2

2. The chemical of most significance in gaseous transport is

   a. Acetylcholine

   b. Hemoglobin

   c. Adrenaline

   d. Antihistamines

3.–6. Fill in the blanks to complete the following sentences:

   Upon inhalation, molecules of **3.** _____ diffuse into the lung's tissues. From there, these molecules then diffuse into **4.** _____ cells, which contain a pigment called **5.** _____. Simultaneously, a second substance formed during cellular respiration, **6.** _____, is released to the lung tissues to be expelled during exhalation.

7. The average breathing rate per minute of adults is

a. 120 breaths

b. 100 breaths

c. 72 breaths

d. 20 breaths

8.–12. Match the respiration terms with their descriptions.

8. _____ Anoxia

9. _____ Internal respiration

10. _____ Asphyxia

11. _____ Hypoxia

12. _____ Pulmonary, or external, respiration

a. Low O_2 content in air breathed in

b. O_2 lacking and excessive CO_2

c. Gaseous exchange between capillaries and cells

d. Without O_2 at the cell level

e. Gaseous exchange in lungs

Inhaling the Basics about the Respiratory Tract

We fill and empty our lungs by contracting and relaxing the respiratory muscles, which include the dome-shaped diaphragm and the intercostal muscles that surround the rib cage. As these muscles contract, air moves through a series of interconnected chambers in the following order (see Figure 8-2): Nose → Pharynx → Larynx → Trachea → Bronchi → Bronchioles → Alveolar ducts → Alveoli. We look at the details of each chamber in that order.

Knowing about the nose (and sinuses)

You may care a great deal about how your nose is shaped, but the shape actually makes little difference to your body. The nose is simply the most visible part of your respiratory tract. Beyond those oh-so-familiar nostrils — which are formally called *nares* — the *septum* divides the nasal cavity into two chambers called the *nasal fossae*. The internal openings at the posterior of the fossae are called the *choanae*. Inside the nostril is a slight dilation extending to the apex of the nose called the *vestibule;* it's lined by skin covered with hairs, plus mucous glands and sebaceous glands that help trap dust and particles before they can enter the lungs.

Each nasal cavity is divided into an *olfactory region* and a *respiratory region*.

✔ The **olfactory region** lies in the upper part of the nasal cavity. Fine filaments distributed over its mucous membrane are actually special nerves devoted to the sense of smell. The bipolar olfactory cells' axons thread through openings in the *cribriform plate* (from the Latin *cribrum* for "sieve") and then come together to form the olfactory nerve (cranial nerve I) that terminates in the olfactory centers of the brain's cerebral cortex.

✔ The nasal cavity's **respiratory region** is covered by a mucous membrane made up of pseudostratified, ciliated columnar epithelium (remember the ten types of epithelium in Chapter 4?) containing mucous and sebaceous glands. The secretions from these glands form a protective layer that warms, moistens, and helps to filter air as it's inhaled. Beneath the protective layer, areolar connective tissue containing *lymphocytes* (which form a thin *lymphoid tissue*) removes foreign materials. A layer of blood vessels next to the *periosteum* (the membrane covering the surface of bones) forms a rich plexus (network) that tends to swell when irritated or inflamed, closing the *ostia* (openings) of the nasal sinuses. Don't you just hate it when that happens?

Ah, the sinuses. They can be such headaches. Lined with a ciliated columnar epithelium (refer to Chapter 4's tissue discussion), *sinuses* are cavities in the bone that reduce the skull's weight and act as resonators for the voice. Each of the sinuses is named for the bone containing it, as follows:

- ✔ **Frontal sinuses** are located in the front bone behind the eyebrows. (If you've ever flown in an airplane with a sinus infection, these are the suckers that hit you right behind the eyes.)
- ✔ **Maxillary sinuses** are located in the *maxillae,* or upper lip.
- ✔ **Ethmoid** and **sphenoid sinuses** are located in the ethmoid and sphenoid bones in the cranial cavity's floor.

Beyond the sinuses and connected to them are *nasal ducts* that extend from the medial angle of the eyes to the *nasal cavity.* These ducts let serous fluid — a biology term referring to any fluid resembling serum — from the eyes' *lacrimal glands* (tear ducts) flow into the nasal cavity.

The nasal cavity performs several important functions:

- ✔ It drains mucous secretions from the sinuses.
- ✔ It drains lacrimal secretions from the eyes.
- ✔ It prepares inhaled air for the lungs by warming, moistening, and filtering it. Dust and bacteria are caught in the mucous and passed outward from the nasal cavity by the motion of the cilia. Some of that gunk is taken up by *lymphatic tissue* in the nasal cavity and respiratory tubes for delivery to the lymph nodes, which destroy invading germs.

Beyond the nasal cavity is the *nasopharynx,* which connects — you guessed it — the nasal cavity to the *pharynx.*

With a bit of a refresher on the nasal and sinus passages, do you think you can hit the following practice questions on the nose?

13. Which of the following statements about the mucous membranes of the nasal cavity is *not* true?

a. They contain an abundant blood supply.

b. They moisten the air that flows over them.

c. They're composed of stratified squamous epithelium.

d. They become inflamed, causing the membrane to swell and close the nasal sinuses.

14. A nasal sinus would not be found in the

a. Frontal bone

b. Ethmoid

c. Maxilla

d. Vomer

e. Sphenoid

15.–30. Use the terms that follow to identify the structures of the respiratory tract shown in Figure 8-2.

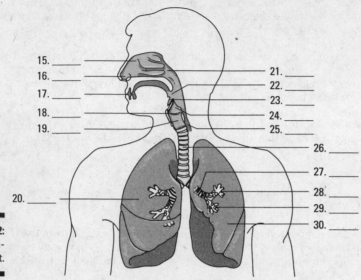

15. _____
16. _____
17. _____
18. _____
19. _____
20. _____
21. _____
22. _____
23. _____
24. _____
25. _____
26. _____
27. _____
28. _____
29. _____
30. _____

Figure 8-2:
The respiratory tract.

Illustration by LifeART Image Copyright © 2007. Wolters Kluwer Health - Lippincott Williams & Wilkins.

a. Esophagus

b. Larynx

c. Nasal cavities

d. Oropharynx

e. Nose

f. Right lung

g. Epiglottis

h. Mouth

i. Alveoli

j. Nasopharynx

k. Thyroid cartilage

l. Left lung

m. Trachea

n. Bronchioles

o. Laryngopharynx

p. Left bronchus

Dealing with throaty matters

In laymen's terms, it's the throat. But you know better, right? The top of the throat consists of these key parts:

✔ **Pharynx:** The pharynx is an oval, fibromuscular sac about 5 inches long and tapering to ½ inch in diameter at its *anteroposterior end,* which is a fancy biology term meaning "front to back." In fact, the point where the pharynx connects to the esophagus is the narrowest part of the entire digestive tract. Eustachian tubes connected to the middle ears enter the pharynx on each side. On the back wall of the pharynx is a mass of lymphoid tissue called the *pharyngeal tonsil,* or *adenoids.*

✔ **Larynx:** Connecting the pharynx with the trachea, this collection of nine cartilages is what makes a man's prominent Adam's apple. Also called the voice box, the larynx looks like a triangular box flattened dorsally and at the sides that becomes narrow and cylindrical toward the base (see Figure 8-3). Ligaments connect the cartilages controlled by several muscles; the inside of the larynx is lined with a mucous membrane that continues into the trachea.

Three of the larynx's nine cartilages go solo — the *thyroid,* the *cricoid,* and the *epiglottis* — while three more come in pairs — the *arytenoids,* the *corniculates,* and the *cuneiforms.* The thyroid cartilage (thyroid in Greek means "shield-shaped) is largest and consists of two plates called *laminae* that are fused just beneath the skin to form a shield-shaped process, the Adam's apple. Immediately above the Adam's apple, the laminae are separated by a V-shaped notch called the superior thyroid notch. The ring-shaped cricoid cartilage is smaller but thicker and stronger, with shallow notches at the top of its broad back that connect, or articulate, with the base of the arytenoid cartilages. The arytenoid cartilages both are shaped like pyramids, with the vocal folds attached at the back and the controlling muscles that move the arytenoids attached at the sides, moving the vocal cords. On top of the arytenoids are the corniculate cartilages, small conical structures for attachment of muscles regulating tension on the vocal cords. Nestled in front of these and inside the *aryepiglottic* fold, the cuneiform cartilages stiffen the soft tissues in the vicinity. The epiglottis, sometimes called the lid on the voice box, is a leaf-shaped cartilage that projects upward behind the root of the tongue. Attached at its stem end, the epiglottis opens during respiration and reflexively closes during swallowing to keep food and liquids from getting into the respiratory tract.

Two types of folds play different roles inside the larynx. The true vocal folds, or cords, are V-shaped when relaxed. When talking, the folds stretch for high sounds or slacken for low sounds, causing the opening into the glottis — the opening in the larynx — to form an oval. Sounds are produced when air is forced over the folds, causing them to vibrate. Just above these folds are the ventricular vocal folds, also known as vestibular or false folds, that don't produce sounds. Muscle fibers within these folds help close the glottis during swallowing.

Following are some practice questions dealing with the throat:

31. The vocal folds change position by the movement of the cartilage known as

a. Cuneiforms

b. Thyroid

c. Arytenoids

d. Cricoid

e. Epiglottis

32.–36. Match the anatomical structure with its function.

32. _____ Epiglottis   **a.** Voice box lid

33. _____ Nasal fossae   **b.** Respiratory center

34. _____ Medulla oblongata   **c.** Prevents collapse of trachea

35. _____ C-shaped cartilaginous rings   **d.** Nasal cavity

36. _____ Thyroid cartilage   **e.** Adam's apple

37. The opening between the two vocal folds is called the

  a. Epiglottis

  b. Bronchi

  c. Alveoli

  d. Glottis

  e. Larynx

38. The loud voice of a person speaking is due to

  a. Vibrating vocal folds

  b. Vibrating chest muscles

  c. Increased air from lungs

  d. Vibrating pharynx

  e. Vibrating trachea

39.–52. Use the terms that follow to identify the structures of the larynx shown in Figure 8-3. Some terms may be used more than once.

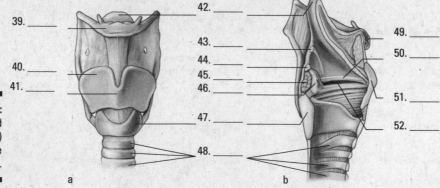

Figure 8-3:
Front (a) and lateral (b) views of the larynx.

Illustration by Imageering Media Services Inc.

  a. Thyroid cartilage

  b. Cricoid cartilage

  c. Hyoid bone

  d. Epiglottis

    e. Arytenoid cartilage

    f. Ventricular fold (false vocal cord)

    g. Laryngenal prominence (Adam's apple)

    h. Cuneiform cartilage

    i. Arytenoid muscle

    j. True vocal cord

    k. Tracheal cartilages

    l. Corniculate cartilage

Going deep inside the lungs

After the pharynx and larynx comes the *trachea,* more popularly known as the wind-pipe. Roughly 6 inches long in adults, it's a tube connected to the larynx in front of the esophagus that's made up of C-shaped rings of hyaline cartilage and fibrous connective tissue that strengthen it and keep it open. Like the larynx, the trachea's lined with mucous membrane covered in cilia. Just above the heart, the trachea splits into two *bronchi* divided by a sharp ridge called the *carina,* with each leading to a lung. But they're not identical: The right *primary bronchus* is shorter and wider than the left primary bronchus. Each primary bronchus divides into secondary bronchi with a branch going to each lobe of the lung; the right side gets three secondary bronchi while the left gets only two. Once inside a designated lobe, the bronchus divides again into tertiary bronchi. The right lung has ten such branches: three in the superior (or upper) lobe, two in the middle lobe, and five in the inferior (or lower) lobe. The left lung has only four tertiary bronchi: two in the upper lobe and two in the lower lobe.

Each tertiary bronchi subdivides one more time into smaller tubes called *bronchioles* (see Figure 8-4), which lack the supporting cartilage of the larger structures. Each bronchiole ends in an elongated sac called the *atrium* (also known as an *alveolar duct* or *alveolar sac*). Alveoli (or air cells) surround the atria, as do small *capillaries* that pick up oxygen for delivery elsewhere in the body and dump off carbon dioxide fetched from elsewhere. Overall, there are 23 branches in the respiratory system, with a combined surface area (counting the alveoli) the size of a tennis court!

Knowing that the bronchi aren't evenly distributed, you may have guessed that the lungs aren't identical either. You're right. They're both spongy and porous because of the air in the sacs, but the right lung is larger, wider, and shorter than the left lung and has three *lobes.* The left lung divides into only two lobes and is both narrower and longer to make room for the heart because two-thirds of that organ lies to the left of the body's midline. Each lobe is made up of many *lobules,* each with a bronchiole ending in an atrium inside.

Covering each lung is a thin serous membrane called the *visceral pleura* that folds back on itself to form a second outer layer, the *parietal pleura,* with a *pleural cavity* between the two layers. These two layers secrete a watery fluid into the cavity to lubricate the surfaces that rub against each other as you breathe. When the pleural membrane becomes inflamed in a condition called *pleurisy,* a sticky discharge roughens the pleura, causing painful irritation. An accompanying bacterial infection means that pus accumulates in the pleural cavity in a condition known as *empyema.*

Blood comes to the lungs through two sources: the *pulmonary arteries* and the *bronchial arteries*. The pulmonary trunk comes from the right ventricle of the heart and then branches into the two pulmonary arteries carrying *venous blood* (the only arteries that contain blood loaded with carbon dioxide from various parts of the body) to the lungs. That blood goes through capillaries in the lungs where the carbon dioxide leaves the blood and enters the alveoli to be expelled during exhalation; oxygen leaves the alveoli through the capillaries to enter the bloodstream. After that, oxygenated arterial blood returns to the left atrium through the pulmonary veins (the only veins that contain oxygenated blood), completing the cycle. Bronchial arteries branch off the thoracic aorta of the heart, supplying the lung tissue with nutrients and oxygen.

Following are some practice questions dealing with the lungs:

53. Cartilage is not found in the

  a. Primary bronchi

  b. Secondary bronchi

  c. Bronchioles

  d. Trachea

  e. Larynx

54. Gaseous exchange occurs between the capillaries and the

  a. Trachea

  b. Alveolar sacs

  c. Primary bronchi

  d. Terminal bronchioles

  e. Secondary bronchi

55.–59. Fill in the blanks to complete the following sentences:

  The trachea divides into two **55.** _____, which then divide into **56.** _____ with a branch going to each lobe of the lung. Upon entering the lobe, each divides into **57.** _____, subdividing into smaller tubes called **58.** _____. They terminate in an elongated sac called the atrium surrounded by **59.** _____ or air cells.

60. If a pin were to pierce the body from the outside in the thoracic region, the third structure it would reach would be the

  a. Pleural cavity

  b. Visceral pleura

  c. Parietal pleura

  d. Lung

61.–69. Use the terms that follow to identify the structures of the bronchiole shown in Figure 8-4.

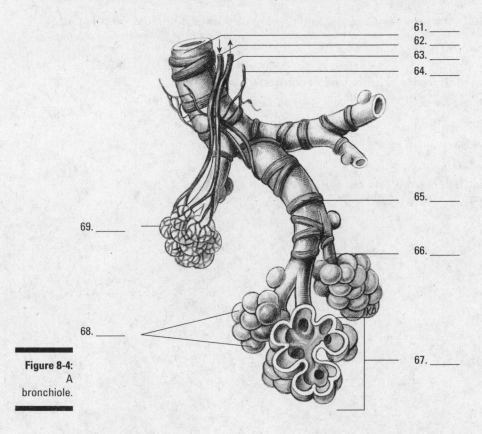

61. _____
62. _____
63. _____
64. _____
65. _____
66. _____
67. _____
68. _____
69. _____

Figure 8-4:
A
bronchiole.

a. Smooth muscle

b. Pulmonary venule

c. Alveolar sac

d. Terminal bronchiole

e. Pulmonary capillary

f. Pulmonary arteriole

g. Alveolar duct

h. Lymphatic vessel

i. Alveoli

Damaging Air

A number of pulmonary diseases can plague human lungs. Inhaling metal and mineral dust can be particularly harmful because the particles cut into and embed themselves in delicate lung tissue, leaving nonfunctional and less pliable scar tissue. Specific lung conditions include

✔ **Silicosis,** commonly found among construction workers, is caused by deposits of sand particles in the lungs.

> ✔ **Anthracosis,** or black lung, occurs among coal miners because of coal dust accumulating in the lungs.
>
> ✔ **Rhinitis,** or the common cold, can be caused by several different kinds of viral infections. Undue exposure may activate the virus or cause the body to become more susceptible to the virus.

70. The disease referred to as anthracosis is caused by

   a. A bacterial infection

   b. Inhaling coal dust

   c. Inhaling sand particles

   d. Undue exposure

   e. Inflammation of the pleura membrane

Answers to Questions on the Respiratory System

The following are answers to the practice questions presented in this chapter.

1 Which of the following gases are dissolved and held in chemical combination in the blood? **c. Oxygen and CO_2.** Yes, the human body's two greatest needs are to inhale oxygen and to exhale carbon dioxide.

2 The chemical of most significance in gaseous transport is **b. hemoglobin.**

Remember that the Latin root for blood is *hemo;* none of the other answer options incorporate that root.

3–6 Upon inhalation, molecules of **3. oxygen** diffuse into the lung's tissues. From there, these molecules then diffuse into **4. red blood** cells, which contain a pigment called **5. hemoglobin.** Simultaneously, a second substance formed during cellular respiration, **6. carbon dioxide**, is released to the lung tissues to be expelled during exhalation.

7 The average breathing rate per minute of adults is **d. 20 breaths.** The rates in the other answer options are more akin to pulse rates or blood pressures than to average respirations per minute.

8 Anoxia: **d. Without O_2 at the cell level**

9 Internal respiration: **c. Gaseous exchange between capillaries and cells**

10 Asphyxia: **b. O_2 lacking and excessive CO_2**

11 Hypoxia: **a. Low O_2 content in air breathed in**

12 Pulmonary, or external, respiration: **e. Gaseous exchange in lungs**

13 Which of the following statements about the mucous membranes of the nasal cavity is *not* true? **c. They're composed of stratified squamous epithelium.**

14 A nasal sinus would not be found in the **d. vomer.**

15–30 Following is how Figure 8-2, the respiratory tract, should be labeled.

15. **c. Nasal cavities**; 16. **e. Nose**; 17. **h. Mouth**; 18. **k. Thyroid cartilage**; 19. **b. Larynx**; 20 **f. Right lung**; 21. **j. Nasopharynx**; 22. **d. Oropharynx**; 23. **g. Epiglottis**; 24. **o. Laryngopharynx**; 25. **a. Esophagus**; 26. **m. Trachea**; 27. **p. Left bronchus**; 28. **n. Bronchioles**; 29. **i. Alveoli**; 30. **l. Left lung**

31 The vocal folds change position by the movement of the cartilage known as **c. arytenoids.** This cartilage is tough to spell and pronounce but easy to move.

32 Epiglottis: **a. Voice box lid**

33 Nasal fossae: **d. Nasal cavity**

34 Medulla oblongata: **b. Respiratory center**

35 C-shaped cartilaginous rings: **c. Prevents collapse of trachea**

36 Thyroid cartilage: **e. Adam's apple**

37 The opening between the two vocal folds is called the **d. glottis.**

38 The loud voice of a person speaking is due to **c. increased air from lungs.** That's why people tend to take a deep breath before they start yelling.

39 - 52 Following is how Figure 8-3, the larynx, should be labeled.

> 39. **c. Hyoid bone;** 40. **a. Thyroid cartilage;** 41. **g. Laryngenal prominence (Adam's apple);** 42. **d. Epiglottis;** 43. **h. Cuneiform cartilage;** 44. **l. Corniculate cartilage;** 45. **e. Arytenoid cartilage;** 46. **i. Arytenoid muscle;** 47. **b. Cricoid cartilage;** 48. **k. Tracheal cartilages;** 49. **c. Hyoid bone;** 50. **f. Ventricular fold (false vocal cord);** 51. **a. Thyroid cartilage;** 52. **j. True vocal cord**

53 Cartilage is not found in the **c. bronchioles.** They're so small that they need to be more elastic and less cartilaginous.

54 Gaseous exchange occurs between the capillaries and the **b. alveolar sacs.** These sacs are the smallest parts of the lungs, so it makes sense that molecular exchange would take place here.

55 - 59 The trachea divides into two **55. primary bronchi,** which then divide into **56. secondary bronchi** with a branch going to each lobe of the lung. Upon entering the lobe, each divides into **57. tertiary bronchi,** subdividing into smaller tubes called **58. bronchioles.** They terminate in an elongated sac called the atrium surrounded by **59. alveoli** or air cells.

60 If a pin were to pierce the body from the outside in the thoracic region, the third structure it would reach would be the **b. visceral pleura.** Note that the question asks you to choose from the list provided, not from the entire structure of the body.

61 - 69 Following is how Figure 8-4, the bronchiole, should be labeled.

> 61. **d. Terminal bronchiole;** 62. **f. Pulmonary arteriole;** 63. **b. Pulmonary venule;** 64. **h. Lymphatic vessel;** 65. **a. Smooth muscle;** 66. **g. Alveolar duct;** 67. **c. Alveolar sac;** 68. **i. Alveoli;** 69. **e. Pulmonary capillary**

70 The disease referred to as anthracosis is caused by **b. inhaling coal dust.** Anthracosis equals black lung equals coal dust.

Chapter 9

Fueling the Functions: The Digestive System

It's time to feed your hunger for knowledge about how nutrients fuel the whole package that is the human body. In this chapter, we help you swallow the basics about getting food into the system and digest the details about how nutrients move into the rest of the body. You also get plenty of practice following the nutritional trail from first bite to final elimination.

Digesting the Basics: It's Alimentary!

Before jumping into a discussion on the alimentary tract, we need to review some basic terms.

✔ **Ingestion:** Taking in food

✔ **Digestion:** Changing the composition of food — splitting large molecules into smaller ones — to make it usable by the cells

✔ **Deglutition:** Swallowing, or moving food from the mouth to the stomach

✔ **Absorption:** Occurs when digested food moves through the intestinal wall and into the blood

✔ **Egestion:** Eliminating waste materials or undigested foods at the lower end of the digestive tract; also known as *defecation*

The alimentary tract develops early on in a growing embryo. The primitive gut, or *archenteron,* develops from the *endoderm* (inner germinal layer) during the third week after conception, a stage during which the embryo is known as a *gastrula*. At the anterior end (head end), the oral cavity, nasal passages, and salivary glands develop from a small depression called a *stomodaeum* in the *ectoderm* (outer germinal layer). The anal and urogenital structures develop at the opposite, or posterior, end from a depression in the ectoderm called the *proctodaeum*. In other words, the digestive tract develops from an endodermal tube with ectoderm at each end.

Whereas the respiratory tract is a two-way street — oxygen flows in and carbon dioxide flows out — the digestive tract is designed to have a one-way flow (although when you're sick or your body detects something bad in the food you've eaten, what goes down sometimes comes back up). Under normal conditions, food moves through your body in the following order (see Figure 9-1):

Mouth → Pharynx → Esophagus → Stomach → Small intestine → Large intestine

When you swallow food, it's mixed with digestive enzymes in both saliva and stomach acids. Circular muscles on the inside of the tract and long muscles along the outside of the tract keep the material moving right through defecation at the end of the line.

Chew on these sample questions about the alimentary tract:

1.–5. Match the alimentary tract terms with their descriptions.

1. _____ Taking in food   **a.** Digestion
2. _____ Elimination of waste   **b.** Ingestion
3. _____ Movement of food from mouth to stomach   **c.** Deglutition
4. _____ Means of transporting food into the blood   **d.** Absorption
5. _____ Mechanical/chemical changing of food composition   **e.** Egestion

6.–16. Use the terms that follow to identify the parts of the digestive system shown in Figure 9-1.

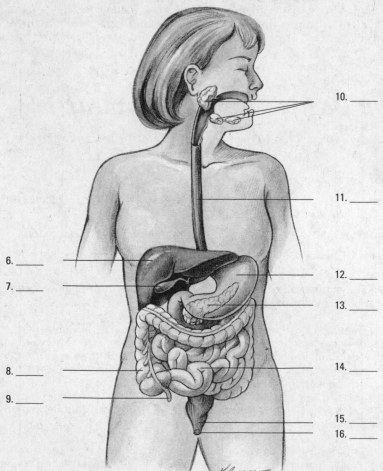

6. _____

7. _____

8. _____

9. _____

10. _____

11. _____

12. _____

13. _____

14. _____

15. _____

16. _____

Figure 9-1:
The organs and glands of the diges-tive system.

    a. Pancreas

    b. Colon

    c. Liver

    d. Small intestine

    e. Salivary glands

    f. Gallbladder

    g. Appendix

    h. Anus

    i. Esophagus

    j. Rectum

    k. Stomach

17. The alimentary tract forms from the following layer(s) of the developing embryo:

    a. Endoderm

    b. Ectoderm

    c. Both the endoderm and the ectoderm

    d. Neither the endoderm nor the ectoderm

18. Identify the correct sequence of the movement of food through the body:

    a. Mouth → Pharynx → Esophagus → Stomach → Small intestine → Large intestine

    b. Mouth → Esophagus → Pharynx → Stomach → Small intestine → Large intestine

    c. Mouth → Pharynx → Esophagus → Stomach → Large intestine → Small intestine

    d. Mouth → Pharynx → Stomach → Esophagus → Small intestine → Large intestine

Nothing to Spit At: Into the Mouth and Past the Teeth

In addition to being very useful for communicating, the mouth serves a number of important roles in the digestive process:

- Chewing, formally known as *mastication,* breaks down food mechanically into smaller particles.

- The act of chewing increases blood flow to all the mouth's structures and the lower part of the head.

- Saliva from *salivary glands* in the mouth helps prepare food to be swallowed and begins the chemical breakdown of carbohydrates.

- Taste buds on the tongue stimulate saliva production. Interestingly, studies have shown that taste preferences can change in reaction to the body's specific needs. In addition, the smell of food can get gastric juices flowing in preparation for digestion.

The mouth's anatomy begins, of course, with the lips, which are covered by a thin, modified mucous membrane. That membrane is so thin that you can see the red blood in the underlying capillaries. (That's the unromantic reason for the lips' natural rosy glow.) The mouth itself is divided into two regions defined by the arches of the upper and lower jaws. The *vestibule* is the region between these *dental arches,* cheeks, and lips, whereas the *oral cavity* is the region inside the dental arches.

Entering the vestibule

The inner surface of the lips is covered by a mucous membrane. Sickle-shaped pieces of tissue called *labial frenula* attach the lips to the gums. Within the mucous membrane are *labial glands,* which produce mucus to prevent friction between the lips and the teeth. The cheeks are made up of *buccinator muscles* and a *buccal pad,* a subcutaneous layer of fat. The buccinator muscles keep the food between the teeth during the act of chewing. Elastic tissue in the mucous membrane keeps the lining of the cheeks from forming folds that would be bitten during chewing (usually — most people have bitten the insides of their cheeks at one time or another). Also stashed away in the cheek, just in front of and below each ear, is a *parotid gland,* which is the largest salivary gland; it releases saliva through a duct opposite the second upper molar tooth. Two other pairs of salivary glands also secrete into the mouth: the *submaxillary glands* along the side of the lower jaw and the *sublingual glands* in the floor of the mouth near the chin.

The dental arches are formed by the *maxillae* (upper jaw) and the *mandible* (lower jaw) along with the *gingivae* (gums) and teeth of both jaws. The gingivae are dense, fibrous tissues attached to the teeth and the underlying jaw bones; they're covered by a mucous membrane extending from the lips and cheeks to form a collar around the neck of each tooth. The gums are very vascular (meaning that lots of blood vessels run through them) but poorly innervated (meaning that, fortunately, they're not generally very sensitive to pain).

Teeth rise from openings in the jawbone called *sockets,* or *alveoli.* You have a number of different kinds of teeth, and each has a specific contribution to the process of biting and chewing. Humans get two sets of teeth in a lifetime. The first temporary, or *deciduous,* set is known as *milk teeth.* Babies between 6 months and 2 years old "cut," or *erupt,* four incisors, two canines, and two molars in each jaw. These teeth are slowly replaced by permanent teeth from about 5 or 6 years of age until the final molars — referred to as *wisdom teeth* — erupt between 17 and 25 years of age.

An adult human has the following 16 teeth in each jaw (for a total set of 32 teeth):

- Four *incisors,* which are chisel-shaped teeth at the front of the jaw for biting into and cutting food

- Two *canines,* or *cuspids,* which are pointed teeth on either side of the incisors for grasping and tearing

- Four *premolars,* or *bicuspids,* which are flatter, shallower teeth that come in pairs just behind the canines

- Six *molars,* which are triplets of broad, flat teeth on either side of the jawbone for grinding and mixing food prior to swallowing

Regardless of type, each tooth has three primary parts, which you can see in Figure 9-2:

- **Crown:** The part that projects above the gum
- **Neck:** The region where the gum attaches to the tooth
- **Root:** The internal structure that firmly fixes the tooth in the alveolus (socket)

Teeth primarily consist of yellowish *dentin* with a layer of *enamel* over the crown and a layer of *cementum* over the root and neck, which are connected to the bone by the *periodontal membrane.* Cementum and dentin are nearly identical in composition to bone; enamel consists of 94 percent calcium phosphate and calcium carbonate and is thickest over the chewing surface of the tooth.

Depending on the structure of the tooth, the root can be a single-, double-, or even triple-pointed structure. In addition, each tooth has a *pulp cavity* at the center that's filled with connective and lymphatic tissue, nerves, and blood vessels that enter the tooth through the root canal via an opening at the bottom called the *apical foramen.* Now you know why it hurts so much when dentists have to drill down and take out that part of an infected tooth!

Moving along the oral cavity

The roof of the oral cavity is formed by both the *hard palate,* a bony structure covered by fibrous tissue and the ever-present mucous membrane, and the *soft palate,* a movable partition of fibromuscular tissue that prevents food and liquid from getting into the nasal cavity. (It's also the tissue that sometimes vibrates in sleep, causing a sonorous grating sound referred to as *snoring.*) The soft palate hangs at the back of the oral cavity in two curved folds that form the palatine arches. The *uvula,* a soft conical process (or piece of tissue), hangs in the center between those folds.

Beyond the soft palate, the *palatopharyngeal* (or *pharyngopalatine*) arch curves sharply toward the midline and blends with the wall of the pharynx, ending at the *dorsum* (back) of the tongue. Another structure, the anterior *palatoglossal* (or *glossopalatine*) arch, starts on the surface of the palate at the base of the uvula and continues in a wide curve forward and downward, ending next to the posterior (back) one-third of the tongue. At the base of these arches and between the folds lie the *palatine tonsils* — if a surgeon hasn't removed them because of frequent childhood infections. The *faucial isthmus* or *oropharynx* is the junction between the oral cavity and the pharynx (described in detail in Chapter 8). It opens during swallowing and closes when you move the dorsum of the tongue against the soft palate when breathing.

The tongue

The tongue, which is a tight bundle of interlaced muscles, and its associated mucous membrane form the floor of the oral cavity. Two distinct groups of muscles — *extrinsic* and *intrinsic* — are used in tandem for mastication (chewing), deglutition (swallowing), and to articulate speech.

✔ The extrinsic muscles, which are used to move the tongue in different directions, originate outside the tongue and are attached to the mandible, styloid processes of the temporal bone and the hyoid and, along with a fold of mucous membrane called the *lingual frenulum,* anchor the tongue.

✔ The intrinsic muscles are a complex muscle network allowing the tongue to change shape for talking, chewing and swallowing.

Three primary types of *papillae* (nipple-shaped protrusions) cover the tongue's forward upper surface:

✔ **Filiform papillae** are fine, brush-like papillae that cover the dorsum, the tip, and the lateral margins of the tongue. They're the most numerous papillae and don't hold any taste buds.

> ✔ **Fungiform papillae** are large, red, mushroom-shaped papillae scattered among the filiform papillae. They have taste buds, which are special receptors that communicate taste signals to the brain.

> ✔ **Vallate papillae,** also called *circumvallate papillae,* are flattened structures, each with a moat-like trough ringing it. There are 12 of these on the tongue, and they surround a V-shaped furrow toward the back of the tongue called the *sulcus terminalis.*

There are no papillae on the back (posterior) one-third of the tongue; that part has only a mucous membrane covering lymphatic tissue, which forms the *lingual tonsils.*

The salivary glands

As we explain in the earlier section "Entering the vestibule," the oral cavity has three pairs of salivary glands producing saliva. The *submandibular* (or *submaxillary*) salivary glands are about the size of a walnut and release fluid onto the floor of the mouth, under the tongue. The smallest pair of the trio, the *sublingual* salivary glands, lies near the tongue under the oral cavity's mucous membrane floor to release secretions directly onto the mucous membrane.

And those secretions are nothing to spit at. Saliva does the following:

✔ Dissolves and lubricates food to make it easier to swallow

✔ Contains *ptyalin,* or *salivary amylase,* an enzyme that initiates chemical digestion of certain carbohydrates

✔ Moistens and lubricates the mouth and lips, keeping them pliable and resilient for speech and chewing

✔ Frees the mouth and teeth of food, foreign particles, and epithelial cells

✔ Produces the sensation of thirst to prevent you from becoming dehydrated

Following are some practice questions regarding the vestibule and oral cavity:

Q. The function of the mouth is

   a. Mixing of solid foods with saliva

   b. Breaking down of the milk protein by the enzyme rennin

   c. Mastication or the breaking down of food into small particles

   d. A and c

   e. A, b, and c

A. The correct answer is mixing of solid foods with saliva and mastication. The mouth does lots of things, including mixing saliva into the food to add the enzyme ptyalin, but that's not rennin. With answer options like these, it's best to stick to the basics.

19.–30. Use the terms that follow to identify the parts of a tooth shown in Figure 9-2.

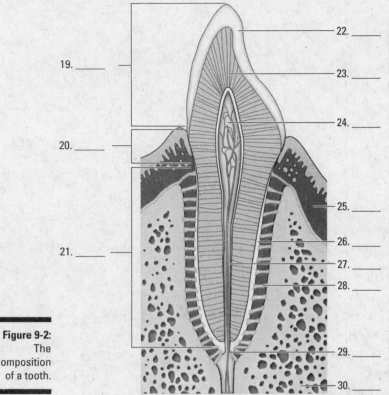

19. _____
20. _____
21. _____

22. _____
23. _____
24. _____
25. _____
26. _____
27. _____
28. _____
29. _____
30. _____

Figure 9-2:
The
composition
of a tooth.

Illustration by Imagineering Media Services Inc.

  a. Root canal

  b. Neck

  c. Bone

  d. Dentin

  e. Crown

  f. Periodontal ligament

  g. Gingiva

  h. Enamel

  i. Root

  j. Apical foramen

  k. Pulp cavity

  l. Cementum

31. The space within the cheek and lip external to the teeth is called the

  a. Rugae

  b. Villi

  c. Fundus

  d. Vestibule

  e. Pylorus

32. The roof of the oral cavity is formed by

    a. The hard and soft palates

    b. The sulcus terminalis

    c. A rigid bony structure covered by fibrous tissue and a mobile partition composed of fibro-muscular tissue in a fold of mucous membrane

    d. A and c

    e. A, b, and c

33. Which of the following statements is *not* true of the teeth?

    a. The permanent teeth in each human jaw are four incisors, two canines, four premolars, and six molars.

    b. Each tooth has a single cuspid anchoring it.

    c. The tooth cavity contains the tooth pulp.

    d. The enamel consists of 94 percent calcium phosphate and calcium carbonate.

    e. Each tooth is composed of a crown, a neck, and a root.

34.–38. Match each description with the proper anatomical structure.

    34. _____ Soft conical process projecting from the soft palate

    35. _____ The junction between the mouth and pharyn

    36. _____ Forms a collar around the teeth and is poorly innervated

    37. _____ Sharply curved arch that bends laterally with the walls of the pharynx

    38. _____ Arch that starts at the buccal surface of the palate at the base of the uvula and ends alongside the back third of the tongue

    a. Pharyngopalatine arch

    b. Faucial isthmus

    c. Gingivae

    d. Glossopalatine arch

    e. Uvula

39. The palatine tonsil is located

    a. In the posterior wall of the pharynx

    b. In the smooth posterior one-third of the tongue

    c. In the region between the rigid hard palate and the soft palate

    d. Under the mucous membrane of the tongue

    e. In the region between the palatopharyngeal and palatoglossal arches

40. The function of saliva is

    a. To facilitate swallowing

    b. To initiate the digestion of certain carbohydrates

    c. To moisten and lubricate the mouth and lips

    d. A and c

    e. A, b, and c

41. The largest salivary gland is the

    a. Submandibular gland

    b. Brunner's gland

    c. Sublingual gland

    d. Submaxillary gland

    e. Parotid gland

42.–44. Match the descriptions with the anatomical structures.

    42. _____ Fine brush-like structures found covering the dorsum of the tongue

    43. _____ Large mushroom-shaped structures

    44. _____ Large structures, each surrounded by a moat that form a V-shaped furrow in the tongue

    a. Vallate papillae

    b. Filiform papillae

    c. Fungiform papillae

Stomaching the Body's Fuel

Deglutition (swallowing) occurs in three phases:

1. **The tip of the tongue elevates slightly, pushing against the hard palate, sliding food onto the back of the tongue, and ultimately propelling it toward the pharynx.**

2. *Tensor muscles* **tighten the palate while** *levator muscles* **raise it until the palate meets the pharyngeal wall, sealing off the nasopharynx from the oropharynx.**

   This action momentarily stops breathing and ensures that food and fluid won't regurgitate through the nose — unless someone makes you laugh, of course.

3. **The** *bolus* **(food mass) heads "down the hatch."**

The pharynx is an oval fibrous muscular sac, about 5 inches long. It opens into the nasal cavity, the oral cavity, the larynx, and the esophagus. On the lateral walls are located the openings to the Eustachian tubes, which connect with the middle ear. In the posterior wall is a mass of lymphatic tissue, the pharyngeal tonsil or adenoid.

This "hatch," borrowed nautical slang for the *esophagus*, is approximately 10 inches long and ½ inch in diameter and carries food through three body regions: the neck, the thorax, and the abdomen. It's not a straight tube, but rather curves slightly to the left as it passes through the diaphragm 1 inch to the left of the midline. The very thick walls of the esophagus are lined with non-keratinized stratified squamous epithelium and include a fibrous outer layer made up of elastic fibers that permit *distention* during swallowing (think of a snake swallowing a whole egg — there's some major stretching going on there). A muscular layer contains both longitudinal and circular layers of smooth muscle. The circular layers contract in sequence, like a series of shrinking and expanding rings, in a movement called *peristalsis* that forces the bolus downward. The longitudinal layers act in concert with the circular muscles, pulling the esophagus over the bolus as it moves downward.

All this pushing and pulling ultimately releases the bolus into the stomach, a pear-shaped bag of an organ that lies just beneath the ribs and diaphragm. About 1 foot long and ½ foot wide, a human stomach's normal capacity is about 1 quart. When empty, the stomach's mucous lining lies in folds called *rugae;* rugae allow expansion of the tummy when you gorge and then shrink the stomach when it's empty to decrease the surface area exposed to acid. Food enters the upper end of the stomach, called the *cardiac* region, through a ring of muscles called the *cardiac sphincter,* which generally remains closed to prevent gastric acids from moving up into the esophagus. The dome-shaped area below the cardiac region is called the *fundus region;* it expands superiorly with really big meals. The lower part of the stomach, shaped like the letter J, is the *pylorus.* The middle part of the *body* of the stomach forms a large curve called the *greater curvature.* The right, much shorter, border of the stomach's body is called the *lesser curvature.* The far end of the stomach remains closed off by the *pyloric sphincter* until its contents have been digested sufficiently to pass into the *duodenum* of the small intestine.

The wall of the stomach consists of three layers of smooth muscle lined by mucous membrane and covered by the peritoneum (see Figure 9-3). The fibers of the outer layer of muscle run longitudinally, the middle layer of muscle consists of circular fibers that encircle the stomach, and the inner layer of muscle fibers runs obliquely only along the fundus region. The stomach's mucous membrane is covered with nonciliated columnar epithelium containing mucous glands.

The three types of gastric glands in the stomach's *epithelium* (lining) are

- **Cardiac glands:** Found in the cardiac region (of course)
- **Pyloric glands:** Secrete mucous in the pyloric region
- **Fundic glands:** Lined with chief cells and parietal cells and are located throughout the stomach's body and fundus

The three types of cells in the *mucosa* (lining) of the stomach are

- **Mucous cells:** Secrete *mucin* (mucous) to protect the mucosa from the high acidity of the gastric juices
- **Chief cells:** Secrete *pepsinogen,* a precursor to the enzyme pepsin that helps break down certain proteins into peptides. (Chief cells in children also produce an enzyme called *rennin,* not found in adults, which acts upon milk proteins.)
- **Parietal cells:** Lie alongside chief cells and secrete the hydrochloric acid that combines with pepsinogen to form pepsin to catalyze protein digestion

The peristaltic contractions that get the bolus into the stomach aren't limited to the esophagus. Instead, peristalsis continues into the musculature of the stomach and stimulates the release of a hormone called *gastrin.* Within minutes, gastrin triggers secretion of gastric juices that reduce the bolus of food to a thick semiliquid mass called *chyme,* which passes through the pyloric sphincter into the small intestine within one to four hours of the food's consumption.

Gastric juices are thin, colorless fluids with an extremely acid pH that ranges from 1 to 4. The quantity of acid released depends on the amount and type of food being digested.

One more part attached to the stomach that we should mention is the *greater omentum.* This is a peritoneal fold that hangs like an apron from the greater curvature of the stomach all the way down to the transverse colon, covering all the small intestine and most of the large intestine. Also called a *caul* or *velum,* this lining can be laden with fat, particularly in obese people.

45.–52. Use the terms that follow to identify the anatomy of the stomach shown in Figure 9-3.

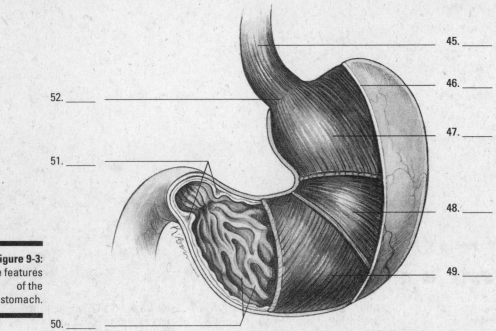

52. _____

51. _____

50. _____

45. _____

46. _____

47. _____

48. _____

49. _____

Figure 9-3:
The features
of the
stomach.

a. Circular muscle layer

b. Esophagus

c. Rugae of the mucosa

d. Cardiac sphincter

e. Serosa

f. Oblique muscle layer

g. Pyloric sphincter

h. Longitudinal muscle layer

53. The sequential contraction of circular muscles as food moves through the esophagus is called

a. Perispasmic contractions

b. Periprostatic contractions

c. Fibrillation

d. Peristalsis

e. Rugae

54. Two muscular rings control movement of food into and out of the stomach. They're called

a. Enzymes

b. Intestines

c. Sphincters

d. Fundic glands

e. Rugae

55. The lower part of the stomach that's shaped like a J is called the

    a. Esophagus

    b. Pylorus

    c. Peritoneal fold

    d. Cardiac region

    e. Fundus region

56. Food that's ready to leave the stomach has been reduced to a thick, semiliquid mass called

    a. Omentum

    b. Gastric juices

    c. Peritoneum

    d. Chyme

    e. Enzymes

Breaking Down the Work of Digestive Enzymes

So what exactly does all the work of digesting and breaking down food? That question brings you back into the realm of proteins. Proteins called *enzymes* act as catalysts, meaning that they initiate and accelerate chemical reactions without themselves being permanently changed in the reaction. Enzymes are very picky proteins indeed; they're effective only in their own pH range, they catalyze only a single chemical reaction, they act on a specific substance called a *substrate,* and they function best at 98.6 degrees Fahrenheit, which just happens to be normal body temperature!

The following sections take you on a tour of the organs in which digestive enzymes do their job.

Small intestine

Most enzyme reactions — in fact most digestion and practically all absorption of nutrients — takes place in the small intestine. Stretching 7 meters (which is nearly 23 feet!), this long snake of an organ extends from the stomach's pylorus to the *ileocecal junction* (the point where the small intestine meets the large intestine), gradually diminishing in diameter along the way.

Three regions of the small intestine play unique roles as chyme moves through it:

    ✔ **Duodenum:** The first section of the small intestine is also the shortest and widest section. As partially digested food enters the duodenum, its acidity stimulates the intestine to secrete the intestinal hormone *enterocrinin,* which controls the secretion of intestinal juices, stimulates the pancreas to secrete its juices, and stimulates the liver to secrete bile. Both the liver and pancreas share a common opening into the duodenum. Lined with large and numerous villi, the duodenum also has *Brunner's glands* that secrete a clear alkaline mucous. The glands are most numerous near the entry to the stomach and decrease in number toward the opposite, or *jejunum,* end.

✔ **Jejunum:** This region of the small intestine also contains villi, but unlike the duo-denum, it has numerous large circular folds at the beginning that decrease in number toward the *ileum* end.

✔ **Ileum:** *Peyer's patches,* which are aggregates of lymph nodes, line this region of the small intestine, becoming largest and most numerous at the *distal* end. The ileum opens into the *cecum* of the large intestine through the *ileocaecal valve*.

A microscopic look at the small intestine reveals circular folds called *plicae circularis,* which project 3 to 10 millimeters into the intestinal lumen and extend anywhere from half to entirely around the tube. These are permanent folds that don't smooth out even when the intestine is distended. Also present are finger-like projections called *villi* that greatly increase the surface area through which the small intestine can absorb nutri-ents. Each villus contains a network of capillaries and a central lymph vessel, or *lacteal,* which contains a milk-white substance called *chyle.* Simple sugars, amino acids, vitamins, minerals, and water are absorbed by the lacteal and combine to form the triglycerides found in the blood. The surface of the villus is simple columnar epithelium (if you can't recall what that means, flip to Chapter 4). Electron microscopy, which can magnify tissues far more than an optical microscope can, reveals that the surface of each villus is further increased by *microvilli.* Peristalsis continues into the small intestine, shortening and lengthening the villi to mix intestinal juices with food and increase absorption. Intestinal glands lie in the depressions between villi, and packed inside these glands are antimicrobial *Paneth cells* within glands called the *crypts of Lieberkühn,* which secrete enzymes that assist pancreatic enzymes.

Intestinal juices contain three types of enzymes:

✔ **Enterokinase** has no enzyme action by itself, but when added to pancreatic juices, it combines with *trypsinogen* to form *trypsin,* which can break down proteins.

✔ **Erepsins,** or **proteolytic enzymes,** don't directly digest proteins but instead complete protein digestion started elsewhere. They split polypeptide bonds, separating amino acids.

✔ **Inverting enzymes** split disaccharides into monosaccharides as follows:

Enzyme	Disaccharide	Monosaccharides
Maltase	Maltose	Glucose + Glucose
Lactase	Lactose	Glucose + Galactose
Sucrase	Sucrose	Glucose + Fructose

Liver

The largest gland in the body, the liver is divided into a large right lobe and a small left lobe by the *falciform ligament,* another peritoneal fold. Two smaller lobes — the *quadrate* and *caudate* lobes — are found on the lower (inferior) and back (posterior) sides of the right lobe. The quadrate lobe surrounds and cushions the *gallbladder,* a pear-shaped structure that stores and concentrates *bile,* which it empties periodically through the *cystic duct* to the *common bile duct* and on into the duodenum during digestion. Bile aids in the digestion and absorption of fats; it consists of bile pigments, bile salts, and cholesterol.

The liver secretes diluted bile through the *hepatic ducts* into the cystic duct and on into the gallbladder. Liver tissue is made up of rows of cuboidal cells separated by microscopic blood spaces called *sinusoids*. Blood from the *interlobular veins and arteries* circulates through the sinusoids with food and oxygen for the liver cells, picking up materials along the way. The blood then enters the intralobular veins, which carry it to the sublobular veins, which empty into the hepatic vein, which leads to the inferior vena cava. Bile secreted from the liver cells is carried by *biliary canaliculi* (bile capillaries) to the bile ducts and then to the hepatic ducts.

Considering the number of vital roles the liver plays, the complexity of that process isn't too surprising. Among the liver's various functions are

✔ Production of blood plasma proteins including *albumin*, antibodies to fend off disease, a blood anticoagulant called *heparin* that prevents clotting, and bile pigments from red blood cells, the yellow pigment *bilirubin,* and the green bile pigment *biliverdin*

✔ Storage of vitamins and minerals as well as glucose in the form of glycogen

✔ Conversion and utilization through enzyme activity of fats, carbohydrates, and proteins

✔ Filtering and removal of nonfunctioning red blood cells, toxins (isolated by *Kupffer cells* in the liver) and waste products from amino acid breakdown, such as urea and ammonia

Unfortunately, a number of serious diseases can damage the liver. The hepatitis virus inflames the gland, and cirrhosis caused by repeated toxic injury (often through alcohol or other substance abuse) destroys Kupffer cells and replaces them with scar tissue. Also, painful gallstones can develop when cholesterol clumps together to form a center around which the gallstone can form.

Pancreas

Equally important. though not as large as the liver, the *pancreas* looks like a roughly 7-inch long, irregularly shaped prism. It has a broad head lodged in the curve of the duodenum. The head is attached to the body of the gland by a slight constriction called the neck, and the opposite end gradually tapers to form a tail. The pancreatic duct extends from the head to the tail, receiving the ducts of various lobules that make up the gland. It generally joins the common bile duct, but some 40 percent of humans have a pancreatic duct and a common bile duct that open separately into the duodenum.

Uniquely, the pancreas is both an *exocrine gland,* meaning that it releases its secretion externally either directly or through a duct, and an *endocrine gland,* meaning that it produces hormonal secretions that pass directly into the bloodstream without using a duct. However, most of the pancreas is devoted to being an exocrine gland secreting pancreatic juices into the duodenum. The endocrine portion of the gland secretes insulin vital to the control of sugar metabolism in the body through small, scattered clumps of cells known as *islets of Langerhans.* Because it contains sodium bicarbonate, pancreatic juice is alkaline, or base, with a pH of 8. Enzymes released by the pancreas act upon all types of foods, making its secretions the most important to digestion. Its enzymes include pancreatic amylase, or carbohydrate enzymes; pancreatic lipase, or fat enzymes; trypsin, or protein enzymes; and nuclease, or nucleic acid enzymes.

The most commonly known pancreatic disease is called diabetes mellitus, or sugar diabetes, which occurs when the islets of Langerhans cease producing insulin. Without insulin, the body can't use sugar, which builds up in the blood and is excreted by the kidneys.

Large intestine

After chyme works its way through the small intestine, it then must move through 5 feet or so of large intestine. The byproduct of the small intestine's work enters at the *ileocaecal valve* and then moves through the following regions of the large intestine:

Cecum → Vermiform appendix → Ascending colon→ Transverse colon → Descending colon → Sigmoid colon → Rectum → Anus

The large intestine is about 3 inches wide at the start and decreases in width all the way to the anus. As the unabsorbed material moves through the large intestine, excess water is reabsorbed, drying out the material. In fact, most of the body's water absorption takes place in the large intestine. Peristaltic movement continues, albeit rather feebly, in the cecum and ascending colon. The large intestine has a longitudinal muscle layer in the form of three bands running from the cecum to the rectum called the *taenia coli*. The large intestine serves no digestive function and secretes only mucus. It has no villi, nor does it have any intestinal glands. Truly, it is the end of the line.

That's a lot of material to digest. See how much you remember:

57. Which of the following terms doesn't belong?

   a. Enterokinase

   b. Maltose

   c. Amylase

   d. Sucrase

   e. Erepsin

58. The parietal cells of the gastric glands secrete

   a. HCl

   b. Pepsinogen

   c. Trypsinogen

   d. Pepsin

   e. Mucous

59. The liver is *least* likely to be involved in

   a. Production of insulin

   b. Production of bile pigments

   c. Storage of vitamins and minerals

   d. Removal of old blood cells

   e. Formation of glycogen

60. The muscle that contracts to prevent gastric juices of the stomach from entering the esophagus is the

   a. Pyloric sphincter

   b. Cardiac sphincter

   c. Ileocecal sphincter

   d. Fundic sphincter

   e. Gastric sphincter

61. The organ in which most digestion occurs is the

   a. Mouth

   b. Stomach

   c. Esophagus

   d. Large intestine

   e. Small intestine

62. The enzyme found in the intestinal juices that activates the pancreatic enzyme into an active enzyme that can break down protein is called

   a. Maltase

   b. Proteolytic enzyme

   c. Erepsin

   d. Inverting enzyme

   e. Enterokinase

63. What structure of the small intestine is composed of a network of capillaries with a central lymph vessel or lacteal, which contains a milky-white substance?

   a. Rugae

   b. Villi

   c. Paneth cells

   d. Islets of Langerhans

   e. Plicae circularis

64. Microscopically, the liver is composed of rows of cuboidal cells with small blood spaces running between the cells called

   a. Sinusoids

   b. Cubisoids

   c. Freakasoids

   d. Rugae

   e. Biliary canaliculi

Answers to Questions on the Digestive Tract

The following are answers to the practice questions presented in this chapter.

1 Taking in food: **b. Ingestion**

2 Elimination of waste: **e. Egestion**

3 Movement of food from mouth to stomach: **c. Deglutition**

4 Means of transporting food into the blood: **d. Absorption**

5 Mechanical/chemical changing of food composition: **a. Digestion**

6–16 Following is how Figure 9-1, the digestive system, should be labeled.

6. **c. Liver**; 7. **f. Gallbladder**; 8. **b. Colon**; 9. **g. Appendix**; 10. **e. Salivary glands**; 11. **i. Esophagus**; 12. **k. Stomach**; 13. **a. Pancreas**; 14. **d. Small intestine**; 15. **j. Rectum**; 16. **h. Anus**

17 The alimentary tract forms from the following layer(s) of the developing embryo: **c. Both the endoderm and the ectoderm.** Keep in mind that the tube that becomes the digestive tract develops from endoderm with ectoderm at each end.

18 Identify the correct sequence of the movement of food through the body: **a. Mouth →Pharynx → Esophagus → Stomach → Small intestine → Large intestine**

Although remembering the sequence M-P-E-S-small-large can be helpful, you can also try this phrase to jog your memory: Most Phones Enable Speeches, from Small to Large.

19–30 Following is how Figure 9-2, the tooth, should be labeled.

19. **e. Crown**; 20. **b. Neck**; 21. **i. Root**; 22. **h. Enamel**; 23. **d. Dentin**; 24. **k. Pulp cavity**; 25. **g. Gingiva**; 26. **l. Cementum**; 27. **a. Root canal**; 28. **f. Periodontal ligament**; 29. **j. Apical foramen**; 30. **c. Bone**

31 The space within the cheek and lip external to the teeth is called the **d. vestibule.** You enter a building through its vestibule, right? That makes it easy to remember the name of the entrance to the mouth, too.

32 The roof of the oral cavity is formed by **d. a and c (the hard and soft palates and a rigid bony structure covered by fibrous tissue and a mobile partition composed of fibromuscular tissue in a fold of mucous membrane).** Admittedly, the latter answer is just a fancy description of the hard and soft palate, but you need to recognize the fancy descriptions along with the common terms.

33 Which of the following statements is *not* true of the teeth? **b. Each tooth has a single cuspid anchoring it.** You can rule out this answer option as false because a cuspid is a type of tooth, so it makes no sense that each tooth would have another type of tooth anchoring it.

34 Soft conical process projecting from the soft palate: **e. Uvula**

35 The junction between the mouth and pharynx: **b. Faucial isthmus**

36 Forms a collar around the teeth and is poorly innervated: **c. Gingivae**

37 Sharply curved arch that bends laterally with the walls of the pharynx: **a. Pharyngopalatine arch**

38 Arch that starts at the buccal surface of the palate at the base of the uvula and ends alongside the back third of the tongue: **d. Glossopalatine arch**

39 The palatine tonsil is located **e. in the region between the palatopharyngeal and palatoglossal arches.**

40 The function of saliva is **e. a, b, and c.** Multifunctional stuff, that saliva. It facilitates swallowing, initiates the digestion of certain carbohydrates, and moistens and lubricates the mouth and lips.

41 The largest salivary gland is the **e. parotid gland.** It lies below and in front of the ear, hence the Greek roots *para–,* meaning "beside," and *ot–,* meaning "ear."

42 Fine brush-like structures found covering the dorsum of the tongue: **b. Filiform papillae**

43 Large mushroom-shaped structures: **c. Fungiform papillae**

44 Large structures, each surrounded by a moat, that form a V-shaped furrow in the tongue: **a. Vallate papillae**

45–52 Following is how Figure 9-3, the stomach, should be labeled.

> 45. **b. Esophagus**; 46. **e. Serosa**; 47. **h. Longitudinal muscle layer**; 48. **a. Circular muscle layer**; 49. **f. Oblique muscle layer**; 50. **c. Rugae of the mucosa**; 51. **g. Pyloric sphincter**; 52. **d. Cardiac sphincter**

53 The sequential contraction of circular muscles as food moves through the esophagus is called **d. peristalsis.**

A bit of Greek may help you remember this term, which comes from the word *peristaltikos,* which means "to wrap around."

54 Two muscular rings control movement of food into and out of the stomach. They're called **c. sphincters.**

55 The lower part of the stomach that's shaped like a J is called the **b. pylorus.** This question calls upon your knowledge of Greek prefixes and suffixes: *pyl–* means "gate," and *–orus* means "guard."

56 Food that's ready to leave the stomach has been reduced to a thick, semiliquid mass called **d. chyme.**

A silly but effective memory tool for this term is this: When food is ready to leave the stomach, it rings a chime.

57 Which of the following terms doesn't belong? **b. Maltose.** This is a sugar, whereas the other answer options are all enzymes.

58 The parietal cells of the gastric glands secrete **a. HCl.** That's chemical shorthand for hydrochloric acid.

59 The liver is *least* likely to be involved in **a. production of insulin.** Insulin production is the job of the pancreas.

60 The muscle that contracts to prevent gastric juices of the stomach from entering the esophagus is the **b. cardiac sphincter.**

To remember this one, keep in mind that the sphincter that serves this purpose is the closest digestive sphincter to the heart.

61 The organ in which most digestion occurs is the **e. small intestine.** It's certainly the longest path for the food to follow! Seeing as the food spends the most time there, it makes sense that it's the site of a lot of digestion.

62 The enzyme found in the intestinal juices that activates the pancreatic enzyme into an active enzyme that can break down protein is called **e. enterokinase.**

It's tricky to remember which of these enzymes is inactive until it combines with something else. You can either try to memorize the function of each enzyme, or you can pick apart the terms. The prefix _entero–_ comes from the Greek word for "intestine." The suffix _–kinase_ stems from the Greek word for "moving." "Moving through the intestine" sounds like a good guess, don't you think?

63 What structure of the small intestine is composed of a network of capillaries with a central lymph vessel or lacteal, which contains a milky-white substance? **b. Villi.** A network of capillaries must be pretty small, and villi are definitely small. Besides, all but one of the other answer options — rugae, islets of Langerhans, and plicae circularis — aren't even in the small intestine.

64 Microscopically, the liver is composed of rows of cuboidal cells with small blood spaces running between the cells called **a. sinusoids**. To help you answer this question, it may help to hark back to Chapter 8's discussion of the nasal sinuses, which we defined as empty spaces.

Chapter 10

Spreading the Love: The Circulatory System

This chapter gets to the heart of the well-oiled human machine to see how its central pump is the hardest-working muscle in the entire body. From a month after you're conceived to the moment of your death, this phenomenal powerhouse pushes a liquid connective tissue — blood — and its precious cargo of oxygen and nutrients to every nook and cranny of the body, and then it keeps things moving to bring carbon dioxide and waste products back out again. In the first seven decades of human life, the heart beats roughly 2.5 billion times. Do the math: How many pulses has your ticker clocked if the average heart keeps up a pace of 72 beats per minute, 100,000 per day, or roughly 36 million per year?

Moving to the Beat of a Pump

Also called the *cardiovascular system*, the circulatory system includes the heart, all blood vessels, and the blood that moves endlessly through it all (see Figure 10-1). It's what's referred to as a *closed double system*; the term "closed" is used for three reasons: because the blood is contained in the heart and its vessels; because the vessels specifically target the blood to the tissues; and because the heart critically regulates blood flow to the tissues. The system is called "double" because there are two distinct circuits and cavities within the heart separated by a wall of muscle called the *septum*. (Each cavity in turn has two chambers called *atria* on top and *ventricles* below). The double circuits are the following:

✔ The *pulmonary circuit* carries blood to and from the lungs for gaseous exchange. Centered in the right side of the heart, this circuit receives blood saturated with carbon dioxide from the veins and pumps it through the *pulmonary artery* (or *trunk*) to capillaries in the lungs, where the carbon dioxide departs the system. That same blood, freshly loaded with oxygen, then returns to the left side of the heart through the *pulmonary veins* where it enters the second circuit.

✔ The *systemic circuit* uses the oxygen-rich blood to maintain a constant internal environment around the body's tissues. From the left side of the heart, the blood moves through the *aorta* to a variety of systemic arteries for distribution throughout the body.

After oxygen is exchanged for carbon dioxide, the blood returns to the pulmonary circuit on the right side of the heart via the *superior* and *inferior venae cavae* (the singular is *vena cava*).

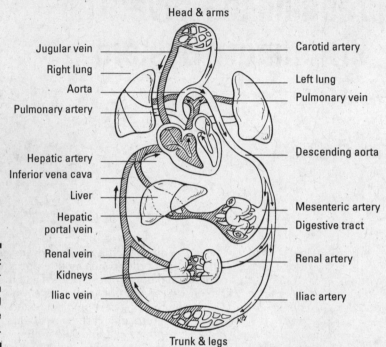

Head & arms

Jugular vein

Carotid artery

Right lung

Left lung

Aorta

Pulmonary artery

Pulmonary vein

Hepatic artery

Descending aorta

Inferior vena cava

Liver

Hepatic
portal vein

Mesenteric artery

Digestive tract

Renal vein

Renal artery

Kidneys

Iliac vein

Iliac artery

Trunk & legs

Figure 10-1:
The circula-
tory system
is a closed
double
system.

Although cutely depicted in popular culture as uniformly curvaceous, the heart actually looks more like a blunt, muscular cone (roughly the size of a fist) resting on the diaphragm. A fluid-filled, fibrous sac called the *pericardium* (or *heart sac*) wraps loosely around the package; it's attached to the large blood vessels emerging from the heart but not to the heart itself. The sternum (breastbone) and third to sixth costal cartilages of the ribs provide protection in front of (ventrally to) the heart. Behind it lie the fifth to eighth thoracic vertebrae. Two-thirds of the heart lies to the left of the body's center, with its *apex* (cone) pointed down and to the left. At less than 5 inches long and a bit more than 3 inches wide, an adult human heart weighs around 10 ounces — a couple ounces shy of a can of soda.

Three layers make up the wall of the heart.

- On the outside lies the *epicardium* (or *visceral pericardium*), which is composed of fibroelastic connective tissue dappled with adipose tissue (fat) that fills external grooves called *sulci* (the singular is *sulcus*). The larger coronary vessels and nerves are found in the adipose tissue that fills the sulci.

- Beneath the epicardium lies the *myocardium,* which is composed of layers and bundles of cardiac muscle tissue.

- The *endocardium*, the heart's interior lining, is composed of simple squamous endothelial cells.

Too much to remember? To keep the layers straight, turn to the Greek roots. *Epi–* is the Greek term for "upon" or "on" whereas *endo–* comes from the Greek *endon* meaning "within." The medical definition of *myo–* is "muscle." And *peri–* comes from the Greek term for "around" or "surround." Hence the *epi*cardium is *on* the heart, the *endo*cardium is *inside* the heart, the *myo*cardium is the *muscle* between the two, and the *peri*cardium surrounds the whole package. By the way, the root *cardi–* comes from the Greek word for heart, *kardia.*

The pericardium is made up of two parts — a tough inelastic sac called the *fibrous pericardium* on the outside and a serous (or lubricated) membrane nearer the heart called the *parietal pericardium.* Between the serous layers of the *epicardium* and the parietal pericardium is the small *pericardial space* and its tiny amount of lubricating *pericardial fluid.* This watery substance prevents irritation during *systole* (contraction of the heart) and *diastole* (relaxation of the heart).

Give these practice questions a try to see if you have the rhythm of all this:

1.–5. Match the description to its anatomical term.

1. _____ The system for gaseous exchange in the lungs

2. _____ The system for maintaining a constant internal environment in other tissues

3. _____ The membranous sac that surrounds the heart

4. _____ The wall that divides the heart into two cavities

5. _____ Uppermost two chambers of the heart

    a. Pericardium

    b. Pulmonary circuit

    c. Systemic circuit

    d. Atria

    e. Septum

6. The heart contracts at an average rate of

    a. 110 times/minute

    b. 72 times/minute

    c. 12 times/minute

    d. 42 times/minute

    e. 24 times/minute

7. A closed system of circulation involves

    a. Non-confinement of blood, general dispersal, minimal regulation

    b. Confinement of blood, general dispersal, critical regulation

    c. Non-confinement of blood, specific targeting, critical regulation

    d. Confinement of blood, specific targeting, critical regulation

    e. Confinement of blood, specific targeting, minimal regulation

8. The inner layer of the heart's wall is called the

    a. Pericardium

    b. Epicardium

    c. Endocardium

    d. Endothelium

9.–13. Match the description to its anatomical term.

9. _____ A membranous, serous layer attached to a fibrous sac

10. _____ A tissue composed of layers and bundles of cardiac muscles

11. _____ Outside layer of the heart wall that's interspersed with adipose

12. _____ The interior lining of the heart

13. _____ External grooves that indicate the regions of the heart

a. Visceral pericardium

b. Sulci

c. Endocardium

d. Myocardium

e. Parietal pericardium

Finding the Key to the Heart's Chambers

The heart's lower two chambers, the ventricles, are quite a bit larger than the two atria up top. Yet the proper anatomical terms for their positions refer to the atria as being "superior" (above) and the ventricles "inferior" (below). In this case, size isn't the issue at all.

The atria

Sometimes referred to as "receiving chambers" because they receive blood returning to the heart through the veins, each atrium has two parts: a *principal cavity* with a smooth interior surface and an *auricle,* a smaller, dog-ear-shaped pouch with muscular ridges inside called *pectinate muscles*, or *musculi pectinati,* that resemble the teeth of a comb.

The right atrium appears slightly larger than the left and has somewhat thinner walls than the left. Its principal cavity, the *sinus venarum cavarum,* is between the two *vena cavae* and the *atrioventricular* (between an atrium and a ventricle) openings. The point where the right atrium's auricle joins with its principal cavity is marked externally by the *sulcus terminalis* and internally by the *crista terminalis.* Openings into the right atrium include the following:

✔ The *superior vena cava,* which has no valve and returns blood from the head, thorax, and upper extremities and directs it toward the atrioventricular opening

✔ The *inferior vena cava,* which returns blood from the trunk and lower extremities and directs it toward the *fossa ovalis* in the interatrial septum, which also has no valve

✔ The *coronary sinus,* which opens between the inferior vena cava and the atrioventricular opening, returns blood from the heart, and is covered by the ineffective *Thebesian valve*

✔ An atrioventricular opening covered by the *tricuspid valve*

> ✔ The *fossa ovalis* is an oval depression in the interatrial septum that corresponds to the foramen ovale of the fetal heart. If the foramen ovale does not close at birth, it causes a condition known as "blue baby."

The left atrium's principal cavity contains openings for the four *pulmonary veins,* two from each lung, which have no valves. Frequently, the two left veins share a common opening. The left auricle, a dog-ear-shaped blind pouch, is longer, narrower, and more curved than the right, marked interiorly by the pectinate muscles, and the left atrium's atrioventricular (or AV) opening is smaller than on the right and is protected by the *mitral,* or *bicuspid,* valve.

The ventricles

The heart's ventricles are sometimes called the pumping chambers because it's their job to receive blood from the atria and pump it back to the lungs and out into the body's network of arteries. More force is needed to move the blood great distances, so the myocardium of the ventricles is thicker than that of either atrium, and the myocardium of the left ventricle is thicker than that of the right.

The right ventricle only has to move blood to the lungs, so its myocardium is only one-third as thick as that of its neighbor to the left. Roughly triangular in shape, the right ventricle occupies much of the *sternocostal* (front) surface of the heart and forms the *conus arteriosus* where it joins the *pulmonary artery,* or *trunk.* The right ventricle extends downward toward where the heart rests against the diaphragm. A circular opening into the pulmonary trunk is covered by the *pulmonary semilunar valve,* so-called because of its three crescent-shaped cusps. When the ventricle relaxes, the blood from the pulmonary artery tends to flow back toward the ventricle, filling the pockets of the cusps and causing the valve to close. The oval AV opening is surrounded by a strong fibrous ring and covered by the *tricuspid valve,* named after its three unequally sized cusps. The atrial surface of the tricuspid valve is smooth, but the side toward the ventricle is irregular, forming a ragged edge where the *chordae tendineae* attach. These fibrous cords, which are attached to nipple-shaped projections called *papillary muscles* in the ventricle's wall, prevent blood from flowing back into the atrium. Cardiac muscle in the ventricle's wall is in an irregular pattern of bundles and bands called the *trabeculae carneae.*

Longer and more conical in shape, the left ventricle's tip forms the apex of the heart. Its walls are three times thicker than those of the right ventricle. Its AV opening is smaller than that in the right ventricle and is covered by the *bicuspid,* or *mitral,* valve that's comprised of two unequal cusps. This ventricle's chordae tendineae are fewer, thicker, and stronger, and they're attached by only two larger papillary muscles, one on the front (anterior) wall and one on the back (posterior). More ridges are packed more densely in the muscular trabeculae carneae. Its opening to the aorta is protected by the *aortic semilunar valve,* composed of three half-moon cusps that are larger, thicker, and stronger than the pulmonary valve's cusps. Between these cusps and the aortic wall are dilated pockets called *aortic sinuses,* which are openings for the *coronary arteries.*

Pump up your practice time with these questions related to the chambers of the heart:

14.–29. Use the terms that follow to identify the heart's major vessels shown in Figure 10-2.

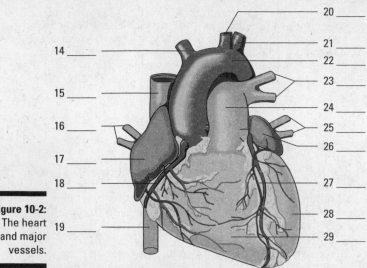

Figure 10-2:
The heart and major vessels.

a. Left pulmonary veins

b. Left ventricle

c. Brachiocephalic trunk

d. Right pulmonary veins

e. Right ventricle

f. Left subclavian artery

g. Right coronary artery

h. Aortic arch

i. Left cardiac vein

j. Superior vena cava

k. Left common carotid artery

l. Left pulmonary arteries

m. Left atrium

n. Inferior vena cava

o. Pulmonary trunk

p. Right atrium

30.–33. Use the terms that follow to identify the heart valves shown in Figure 10-3. (*Note:* In a beating heart, either the two top or the two bottom valves would be open whenever the opposite pair is closed. Figure 10-3, however, gives a better view of all four valves simultaneously.)

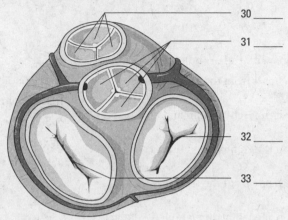

30 _____

31 _____

32 _____

33 _____

Figure 10-3: The heart valves.

LifeART Image Copyright © 2007. Wolters Kluwer Health — Lippincott Williams & Wilkins

    a. Tricuspid valve

    b. Pulmonary semilunar valve

    c. Aortic semilunar valve

    d. Biscuspid valve

34. The cavity in the heart that contains the areas called the sinus venarum cavarum and a blind pouch called the auricle is the

    a. Left ventricle

    b. Right atrium

    c. Left atrium

    d. Right ventricle

35. The superior vena cava enters the heart by way of the

    a. Left ventricle

    b. Pulmonary vein

    c. Right ventricle

    d. Left atrium

    e. Right atrium

36. The cusps of the atrioventricular valves are held in place by

    a. Supporting ligaments

    b. The chordae tendineae

    c. The trabeculae carneae

    d. The papillary muscles

    e. Nothing because they need no supporting structure to hold them

37. The atrioventricular opening between the right atrium and right ventricle is covered by the

 a. Bicuspid valve

 b. Tricuspid valve

 c. Mitral valve

 d. Semilunar valve

38.–42. Match the following descriptions with the proper anatomical terms.

 38. _____ Returns blood to the heart from the head, thorax, and upper extremities

 39. _____ Valve located between the right atrium and right ventricle

 40. _____ Valve located between the right ventricle and pulmonary artery

 41. _____ Returns blood to the heart from the trunk and lower extremities

 42. _____ Valve located between the left atrium and left ventricle

 a. Tricuspid valve

 b. Bicuspid valve

 c. Superior vena cava

 d. Semilunar valve

 e. Inferior vena cava

Conducting the Heart's Music

The mighty, nonstop heart keeps up its rhythm because of a carefully choreographed dance of electrical impulses called the *conduction system* that has the power to produce a spontaneous rhythm and conduct an electrical impulse. Four structures play key roles in this dance — the *sinoatrial node, atrioventricular node, atrioventricular bundle,* and *Purkinje fibers.* (You can see them in Figure 10-4.) Each is formed of highly tuned modified cardiac muscle. Rather than both contracting and conducting impulses as other cardiac muscle does, these structures specialize in conduction alone, setting the pace for the rest of the heart. Following is a bit more information about each one:

 ✔ **Sinoatrial node:** This node really is the pacemaker of the heart. Located at the junction of the superior vena cava and the right atrium, this small knot, or mass, of specialized heart muscle initiates an electrical impulse that moves over the musculature of both atria, causing atrial walls to contract simultaneously and emptying blood into both ventricles. It's also called the *S-A node, sinoauricular node,* and *sinus node.*

 ✔ **Atrioventricular node:** The impulse that starts in the S-A node moves to this mass of modified cardiac tissue that's located in the septal wall of the right atrium. Also called the *A-V node,* it directs the impulse to the A-V bundles in the septum.

 ✔ **Atrioventricular bundle:** From the A-V node, the impulse moves into the atrioventricular bundle, also known as the *A-V bundle* or *bundle of His* (pronounced "hiss"). The bundle breaks into two branches that extend down the sides of the interventricular septum under the endocardium to the heart's apex.

 ✔ **Purkinje fibers:** At the apex, the bundles break up into *terminal conducting fibers,* or Purkinje fibers, and merge with the muscular inner walls of the ventricles. The pulse then stimulates ventricular contraction that begins at the apex and moves toward the base of the heart, forcing blood toward the aorta and pulmonary artery.

One of the best ways to detect cardiac tissue under a microscope is to look for undulating double membranes called *intercalated discs* separating adjacent cardiac muscle fibers. *Gap junctions* in the discs permit ions to pass between the cells, spreading the *action potential* of the electrical impulse and synchronizing cardiac muscle contractions. Potential problems include *fibrillation,* a breakdown in rhythm or propagation of the impulses that causes individual fibers to act independently, and *heart block,* an interruption that causes the atria and ventricles to take on their own rates of contraction. Usually the atria contract faster than the ventricles.

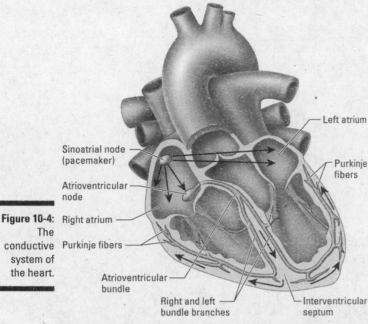

Figure 10-4: The conductive system of the heart.

Left atrium

Sinoatrial node (pacemaker)

Purkinje fibers

Atrioventricular node

Right atrium

Purkinje fibers

Atrioventricular bundle

Right and left bundle branches

Interventricular septum

LifeART Image Copyright © 2007. Wolters Kluwer Health — Lippincott Williams & Wilkins

A healthy heart makes a "lub-dub" sound as it beats. The first sound (the "lub") is heard most clearly near the apex of the heart and comes at the beginning of ventricular systole (the closing of the atrioventricular valves and opening of the semilunar valves). It's lower in pitch and longer in duration than the second sound (the "dub"), heard most clearly over the second rib, which results from the semilunar valves closing during ventricular diastole. Defects in the valves can cause turbulence or regurgitation of blood that can be heard through a stethoscope. Called *murmurs,* these sounds indicate imperfect closure of one or more valves.

Have you got the beat? Try the following practice questions that deal with the heart's rhythm:

Q. Cardiac tissue is distinctive microscopically because of the presence of

    a. Hemoglobin

    b. Intercalated discs

    c. Fibrin

    d. Ganglia

    e. Nuclei

A. The correct answer is intercalated discs.

43. The pacemaker of the heart is also known as the

   a. Sinoatrial node (S-A node)

   b. Extrinsic nerve control center

   c. Bundle of His

   d. Atrioventricular node (A-V node)

   e. Purkinje control fibers

44. Stimulation of myocardial contractions in the ventricles radiates from the

   a. Bundle of His

   b. Atrioventricular bundles

   c. Sinoatrial node (S-A node)

   d. Purkinje fibers

   e. Atrioventricular node (A-V node)

45. Choose the correct conductive pattern.

   a. S-A node → Bundle of His → A-V node → Purkinje fibers

   b. A-V node → S-A node → Purkinje fibers → Bundle of His

   c. S-A node → A-V node → Bundle of His → Purkinje fibers

   d. S-A node → Purkinje fibers → Bundle of His → A-V node

Riding the Network of Blood Vessels

Blood vessels come in three varieties, which you can see illustrated in Figure 10-5:

   ✔ *Arteries* carry blood away from the heart. The largest artery is the aorta. Small ones are called *arterioles,* and microscopically small ones are called *metarterioles.*

   ✔ *Veins* carry blood toward the heart; all veins except the pulmonary veins contain deoxygenated blood. Small ones are called *venules,* and large venous spaces are called *sinuses.*

   ✔ Microscopically small *capillaries* carry blood from arterioles to venules, but sometimes tiny spaces in the liver and elsewhere called *sinusoids* replace capillaries.

The walls of arteries and veins have three layers: the outermost *tunica externa* (sometimes called *tunica adventitia*) composed of white fibrous connective tissue, a central "active" layer called the *tunica media* composed of smooth muscle fibers and yellow elastic fibers, and an inner layer called the *tunica intima* made up of endothelium that aids in preventing blood coagulation by reducing the resistance of blood flow. Arterial walls are very strong, thick, and very elastic to withstand the great pressure to which the arteries are subjected. Arteries have no valves.

There are two types of arteries: elastic and muscular. In elastic arteries, found primarily near the heart, the tunica media is composed of yellow elastic fibers that stretch with each systole and recoil during diastole; essentially they act as shock absorbers to smooth out blood flow. In muscular arteries, the tunica media consists primarily of smooth muscle fibers that are active in blood flow and distribution of blood. The larger blood vessels have smaller blood vessels, the *vasa vasorum,* that carry nourishment to the vessel wall.

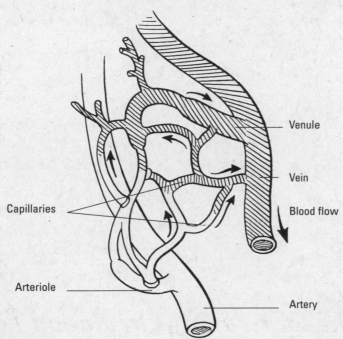

Venule

Vein

Blood flow

Capillaries

Figure 10-5:
The
capillary
exchange.

Arteriole

Artery

While larger in diameter than arteries, veins have thinner walls and are less distensible and elastic. Veins that carry blood against the force of gravity, such as those in the legs and feet, contain valves to prevent backsliding into the capillaries. Normally the blood that veins are returning to the heart is unoxygenated (contains carbon dioxide); the one exception is the pulmonary vein, which returns oxygenated blood to the heart from the lungs.

Capillaries are breathtakingly tiny and capable of forming vast networks, or capillary beds. Their walls are a single layer of squamous endothelial cells. Precapillary sphincters take the place of valves to regulate blood flow. All exchange occurs at the capillaries.

Blood from the digestive tract takes a detour through the hepatic portal vein to the liver before continuing on to the heart. Called the *hepatic portal system*, this circuitous route helps regulate the amount of glucose circulating in the bloodstream (see Figure 10-6). As the blood flows through the sinusoids of the liver, *hepatic parenchymal cells* remove the nutrient materials. *Phagocytic* cells in the sinusoids remove bacteria and other foreign materials from the blood. The blood exits the liver by the hepatic veins, which carry it to the inferior vena cava, which ultimately returns it to the heart.

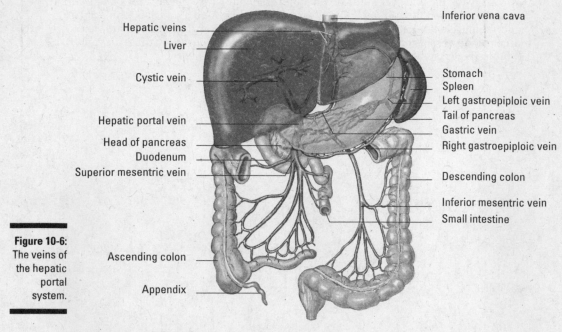

Hepatic veins

Liver

Cystic vein

Hepatic portal vein

Head of pancreas

Duodenum

Superior mesentric vein

Inferior vena cava

Stomach

Spleen

Left gastroepiploic vein

Tail of pancreas

Gastric vein

Right gastroepiploic vein

Descending colon

Inferior mesentric vein

Small intestine

Ascending colon

Appendix

Figure 10-6:
The veins of
the hepatic
portal
system.

Beating from the Start: Fetal Circulation

Because nutrients and oxygen come from the mother's bloodstream, fetal circulation requires extra vessels to get the job done. Two umbilical arteries — the *umbilical vein* and the *ductus venosus* — fill the bill. Fetal blood leaves the placenta through the umbilical vein, which branches at the liver to become the *ductus venosus* before entering the inferior vena cava that carries blood to the right atrium and then through a hole in the septum called the *foramen ovale* into the left atrium. From there it flows into the left ventricle and is pumped through the aorta to the head, neck, and upper extremities. It returns to the heart through the superior vena cava, to the right atrium, to the right ventricle, to the pulmonary trunk (lungs inactive), goes through the ductus arteriosus into the aorta, to the abdominal and pelvic viscera and lower extremities, and to the placenta through the umbilical artery. After birth, these circulation pathways quickly shut down, eventually leaving a depression in the septum, the fossa ovale, where the hole of the foramen ovale once was.

Now is your chance to practice circulating through the circulatory system:

46. In fetal hepatic portal circulation, blood flows directly into the systemic circulation through the

a. Conus arteriosus

b. Hepatic vein

c. Ductus venosus

d. Ductus arteriosus

e. Truncus arteriosus

47. Follow a drop of blood through the heart, starting in the superior vena cava. Number the following structures in sequential order.

_____ Pulmonary vein

_____ Right ventricle

_____ Lung

_____ Right atrium

_____ Pulmonary artery

48. Number the structures in the correct sequence of blood flow from the heart to the radial artery for pulse. Start at the heart with the aortic semilunar valve.

_____ Axillary artery

_____ Subclavian artery

_____ Ascending aorta

_____ Brachial artery

_____ Aortic arch

49. Number the structures in the correct sequence of blood flow from the forearm to the heart.

_____ Basilic vein

_____ Subclavian vein

_____ Superior vena cava

_____ Brachial vein

_____ Axillary vein

50. Number the structures in the correct sequence of blood flow from the great saphenous vein back to the heart.

_____ External iliac vein

_____ Right atrium

_____ Common iliac vein

_____ Femoral vein

_____ Inferior vena cava

51. Follow a drop of blood from the right atrium to the radial artery (for pulse). Number the structures in sequential order.

___1___ Right atrium

_____ Pulmonary artery

_____ Left atrium

_____ Bicuspid valve

_____ Ascending aorta

_____ Axillary artery

_____ Radial artery

_____ Right ventricle

_____ Pulmonary vein

_____ Aortic semilunar valve

_____ Brachial artery

_____ Tricuspid valve

_____ Lung capillary

_____ Aortic arch

_____ Pulmonary semilunar valve

_____ Left ventricle

_____ Subclavian artery

52. Follow a drop of blood from the stomach to the inferior vena cava. (Remember the portal system?) Number the structures in sequential order.

__1__ Superior mesenteric vein

_____ Sinusoids of liver

_____ Hepatic vein

_____ Hepatic portal vein

_____ Inferior vena cava

53. Follow a drop of blood from the aortic semilunar valve of the heart to the forearm and back to the heart. Number the structures in sequential order.

__1__ Ascending aorta

_____ Basilic vein

_____ Axillary artery

_____ Subclavian vein

_____ Brachial vein

_____ Radial artery

_____ Right atrium

_____ Brachial artery

_____ Axillary vein

_____ Capillaries in the hand

_____ Superior vena cava

_____ Subclavian artery

_____ Aortic arch

54. Follow a drop of blood from the saphenous vein back to the heart. Number the structures in sequential order.

__1__ Saphenous vein

_____ External iliac vein

_____ Inferior vena cava

_____ Right atrium

_____ Femoral vein

_____ Common iliac vein

_____ Right ventricle

55. Follow a drop of blood from the anterior tibial vein to the lungs. Number the structures in sequential order.

  _1_   Anterior tibial vein

  _____   External iliac vein

  _____   Inferior vena cava

  _____   Right ventricle

  _____   Popliteal vein

  _____   Common iliac vein

  _____   Pulmonary artery

  _____   Femoral vein

  _____   Right atrium

  _____   Lung capillaries

Answers to Questions on the Circulatory System

The following are answers to the practice questions presented in this chapter.

1   The system for gaseous exchange in the lungs: **b. Pulmonary circuit**

2   The system for maintaining a constant internal environment in other tissues: **c. Systemic circuit**

3   The membranous sac that surrounds the heart: **a. Pericardium**

4   The wall that divides the heart into two cavities: **e. Septum**

5   Uppermost two chambers of the heart: **d. Atria**

6   The heart contracts at an average rate of **b. 72 times/minute.** Faster than that indicates the individual probably is exercising; slower than that means that the individual either is sick or is a highly trained athlete.

7   A closed system of circulation involves **d. confinement of blood, specific targeting, critical regulation.** In short, the system confines, targets, and regulates.

8   The inner layer of the heart's wall is called the **c. endocardium.** Remember that *endo–* is "within," *epi–* is "upon," and *peri–* is "around."

9   A membranous, serous layer attached to a fibrous sac: **e. Parietal pericardium**

10   A tissue composed of layers and bundles of cardiac muscles: **d. Myocardium**

11   Outside layer of the heart wall that's interspersed with adipose: **a. Visceral pericardium**

12   The interior lining of the heart: **c. Endocardium**

13   External grooves that indicate the regions of the heart: **b. Sulci**

14–29   Following is how Figure 10-2, the heart, should be labeled.

14. **c. Brachiocephalic trunk**; 15. **j. Superior vena cava**; 16. **d. Right pulmonary veins**; 17. **p. Right atrium**; 18. **g. Right coronary artery**; 19. **n. Inferior vena cava**; 20. **k. Left common carotid artery**; 21. **f. Left subclavian artery**; 22. **h. Aortic arch**; 23. **l. Left pulmonary arteries**; 24. **o. Pulmonary trunk**; 25. **a. Left pulmonary veins**; 26. **m. Left atrium**; 27. **i. Left cardiac vein**; 28. **b. Left ventricle**; 29. **e. Right ventricle**

30–33   Following is how Figure 10-3, the heart valves, should be labeled.

30. **b. Pulmonary semilunar valve**; 31. **c. Aortic semilunar valve**; 32. **a. Tricuspid valve**; 33. **d. Bicuspid valve**

34   The cavity in the heart that contains the areas called the sinus venarum cavarum and a blind pouch called the auricle is the **b. right atrium.**

35   The superior vena cava enters the heart by way of the **e. right atrium.**

36 The cusps of the atrioventricular valves are held in place by **b. the chordae tendineae.**

37 The atrioventricular opening between the right atrium and right ventricle is covered by the **b. tricuspid valve.** Sorry if this seemed a trick question, but even if you have trouble remembering the heart's right openings from its left ones, you simply need to remember that the bicuspid and the mitral valve are the same thing, so "tricuspid valve" is the only correct answer here.

38 Returns blood to the heart from the head, thorax, and upper extremities: **c. Superior vena cava**

39 Valve located between the right atrium and right ventricle: **a. Tricuspid valve**

40 Valve located between the right ventricle and pulmonary artery: **d. Semilunar valve**

41 Returns blood to the heart from the trunk and lower extremities: **e. Inferior vena cava**

42 Valve located between the left atrium and left ventricle: **b. Bicuspid valve**

43 The pacemaker of the heart is also known as the **a. sinoatrial node (S-A node).**

44 Stimulation of myocardial contractions in the ventricles radiates from the **d. Purkinje fibers.**

45 Choose the correct conductive pattern. **c. S-A node → A-V node → Bundle of His → Purkinje fibers.**

46 In fetal hepatic portal circulation, blood flows directly into the systemic circulation through the **c. ductus venosus.**

47 Follow a drop of blood through the heart, starting in the superior vena cava. Number the structures in sequential order. **1. Right atrium; 2. Right ventricle; 3. Pulmonary artery; 4. Lung; 5. Pulmonary vein**

48 Number the structures in the correct sequence of blood flow from the heart to the radial artery for pulse. Start at the heart with the aortic semilunar valve. **1. Ascending aorta; 2. Aortic arch; 3. Subclavian artery; 4. Axillary artery; 5. Brachial artery**

49 Number the structures in the correct sequence of blood flow from the forearm to the heart. **1. Basilic vein; 2. Brachial vein; 3. Axillary vein; 4. Subclavian vein; 5. Superior vena cava**

50 Number the structures in the correct sequence of blood flow from the great saphenous vein back to the heart. **1. Femoral vein; 2. External iliac vein; 3. Common iliac vein; 4. Inferior vena cava; 5. Right atrium**

51 Follow a drop of blood from the right atrium to the radial artery (for pulse). Number the structures in sequential order. **1. Right atrium; 2. Tricuspid valve; 3. Right ventricle; 4. Pulmonary semilunar valve; 5. Pulmonary artery; 6. Lung capillary; 7. Pulmonary vein; 8. Left atrium; 9. Bicuspid valve; 10. Left ventricle; 11. Aortic semilunar valve; 12. Ascending aorta; 13. Aortic arch; 14. Subclavian artery; 15. Axillary artery; 16. Brachial artery; 17. Radial artery**

52 Follow a drop of blood from the stomach to the inferior vena cava. (Remember the portal system?) Number the structures in sequential order. **1. Superior mesenteric vein; 2. Hepatic portal vein; 3. Sinusoids of liver; 4. Hepatic vein; 5. Inferior vena cava**

53 Follow a drop of blood from the aortic semilunar valve of the heart to the forearm and back to the heart. Number the structures in sequential order. **1. Ascending aorta; 2. Aortic arch;**

3. Subclavian artery; 4. Axillary artery; 5. Brachial artery; 6. Radial artery; 7. Capillaries in the hand; 8. Basilic vein; 9. Brachial vein; 10. Axillary vein; 11. Subclavian vein; 12. Superior vena cava; 13. Right atrium

54 Follow a drop of blood from the saphenous vein back to the heart. Number the structures in sequential order. **1. Saphenous vein; 2. Femoral vein; 3. External iliac vein; 4. Common iliac vein; 5. Inferior vena cava; 6. Right atrium; 7. Right ventricle**

55 Follow a drop of blood from the anterior tibial vein to the lungs. Number the structures in sequential order. **1. Anterior tibial vein; 2. Popliteal vein; 3. Femoral vein; 4. External iliac vein; 5. Common iliac vein; 6. Inferior vena cava; 7. Right atrium; 8. Right ventricle; 9. Pulmonary artery; 10. Lung capillaries**

Chapter 11

Keeping Up Your Defenses:
The Lymphatic System

In This Chapter
▶ Delving into lymphatic ducts
▶ Noodling around with nodes
▶ Exploring the lymphatic organs

You see it every rainy day — water, water everywhere, rushing along gutters and down storm drains into a complex underground system that most would rather not give a second thought. Well, it's time to give hidden drainage systems a second thought: Your body has one. You already know that the body is made up mostly of fluid. *Interstitial* or *extracellular fluid* moves in and around the body's tissues and cells constantly. It leaks out of blood capillaries at the rate of nearly 51 pints a day, carrying various substances to and away from the smallest nooks and crannies. Most of that fluid gets reabsorbed into blood capillaries. But the one or two liters of extra fluid that remain around the tissues become a substance called *lymph* that needs to be managed to maintain fluid balance in the internal environment. That's where the lymphatic system steps in, forming an alternative route for the return of tissue fluid to the bloodstream.

But the lymphatic system is more than a drainage network. It's a body-wide filter that traps and destroys invading microorganisms as part of the body's immune response network. It can remove impurities from the body, help absorb and digest excess fats, and maintain a stable blood volume despite varying environmental stresses. Without it, the cardiovascular system would grind to a halt.

We bet that you won't take your little lymph nodes for granted anymore after you're done with this chapter.

Duct, Duct, Lymph

The story of the lymphatic system (shown in Figure 11-1) begins deep within the body's tissues at the farthest reaches of blood capillaries, where nutrients, plasma, and plasma proteins move out into cells, while waste products like carbon dioxide and the fluid carrying those molecules move back in through a process known as *diffusion*. Roughly 10 percent of the fluid that leaves the capillaries remains deep within the tissues as part of the interstitial (meaning "between the tissues") fluid. But in order for the body to maintain sufficient volume of water within in the circulatory system, eventually this plasma and its protein must get back into the blood. So the lymphatic vessels act as a recycling system to gather, transport, cleanse, and return this fluid to the bloodstream.

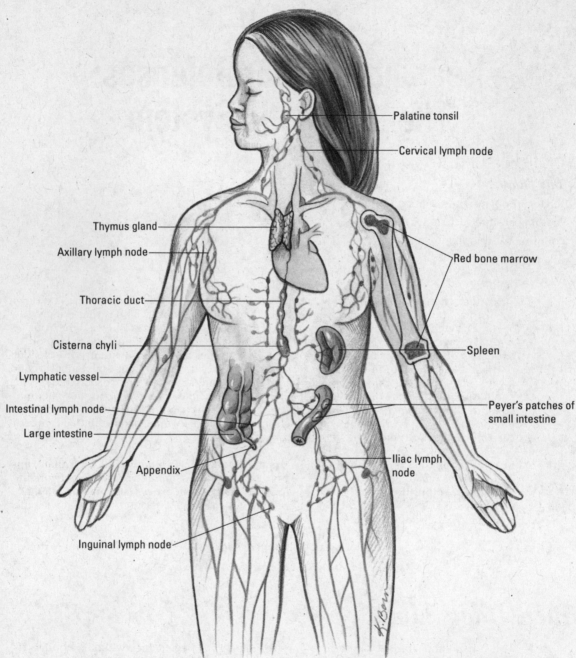

Figure 11-1:
The
lymphatic
system.

Palatine tonsil

Cervical lymph node

Thymus gland

Axillary lymph node

Thoracic duct

Cisterna chyli

Lymphatic vessel

Intestinal lymph node

Large intestine

Appendix

Inguinal lymph node

Red bone marrow

Spleen

Peyer's patches of
small intestine

Iliac lymph
node

To collect the fluid, minute vessels called *lymph capillaries* are woven throughout the body, with a few caveats and exceptions. There are no lymph capillaries in the central nervous system, teeth, outermost layer of the skin, certain types of cartilage, any other avascular tissue, and bones. And because bone marrow makes lymphocytes, which we explain in the next section, it's considered part of the lymphatic system. Plus, *lacteals* (lymphatic capillaries found in the villi of the intestines) absorb fats to mix with lymph, forming a milky fluid called *chyle*. (See Chapter 9 for details on lacteals.) Unlike blood capillaries, lymph capillaries dead-end (terminate) within tissue. Made up of loosely overlapping endothelial cells anchored by fine filaments, lymph capillaries behave as if their walls are made of cellular one-way valves. When the pressure outside the capillary is greater than it is inside, the filaments anchoring the cells allow them to open, permitting interstitial fluid to seep in. Rising differential pressure across the capillary walls eventually forces the cell junctions to close. Once in the capillaries, the trapped fluid is known as *lymph,* and it moves into larger, vein-like *lymphatic vessels.* The lymph moves slowly and without any kind of central pump through a combination of peristalsis, the action of semilunar valves, and the squeezing influence of surrounding skeletal muscles, much like occurs in veins.

In the skin, lymph vessels form networks around veins, but in the trunk of the body and around internal organs, they form networks around arteries. Lymph vessels have thinner walls than veins, are wider, have more valves, and — most important — regularly bulge with bean-shaped sacs called *lymph nodes* (more on those in the later section "Poking at the Nodes"). Just as small tree branches merge into larger ones and then into the trunk, *lymphatics* eventually merge into the nine largest lymphatic vessels called *lymphatic trunks.* The biggest of these at nearly 1½ feet in length is the *thoracic duct;* nearly all the body's lymph vessels empty into it. Only those vessels in the right half of the head, neck, and thorax empty into its smaller mate, the *right lymphatic duct.* Lymph returns to the bloodstream when both ducts connect with the subclavian (under the collarbone) veins.

The thoracic duct, which also sometimes is called the *left lymphatic duct,* arises from a triangular sac called the *chyle cistern* (or *cisterna chyli*) into which one *intestinal trunk* and two *lumbar lymphatic trunks* (which drain the lower limbs) flow. Both the thoracic duct and the much smaller right lymphatic duct drain into the subclavian (behind the collarbone) veins. The remaining four trunks are a pair serving the *jugular* region (sides of the throat) and a pair serving the *bronchomediastinal* region (the central part of the chest).

To see how much of this information is seeping in, answer the following questions:

1. The lymphatic system plays an important role in regulating

   a. Intracellular energy function

   b. Interstitial fluid protein

   c. Metabolizing unused fats from the small intestine

   d. Intercellular transportation of oxygen

   e. None of these

2. Terminated vessels that return plasma proteins to the blood are

   a. Blood capillaries

   b. Venules

   c. Lymph capillaries

   d. Arterioles

   e. Cardiac ducts

3. The thoracic duct does *not* drain lymph from the

   a. Right lower extremity

   b. Right side of the head

   c. Digestive tract

   d. Left axilla

   e. Posterior abdominal wall

4. The largest lymphatic vessel in the body is the

    a. Right lymphatic duct

    b. Spleen

    c. Thoracic duct

    d. Chyle cistern

5. The lymphatic system does *not* function to

    a. Return interstitial fluid to the blood

    b. Destroy bacteria

    c. Remove old erythrocytes

    d. Produce erythrocytes

    e. Produce lymphocytes

Poking at the Nodes

Lymph nodes (see Figure 11-2) are the site of filtration of the lymphatic system. Also sometimes incorrectly referred to as *lymph glands* — they don't secrete anything, so technically they're not glands — these kidney-shaped sacs are surrounded by connective tissue (and therefore are tough to spot). Lymph nodes contain macrophages, which destroy bacteria, cancer cells, and other matter in the lymph fluid. *Lymphocytes*, which produce an immune response to microorganisms, also are found in lymph nodes. Like the kidneys, the indented part of each node is referred to as the *hilus*. The *stroma* (body) of each node is surrounded by a fibrous capsule that dips into the node to form *trabeculae,* or *septa* (thin dividing walls) that divide the node into compartments. *Reticular* (net-like) fibers are attached to the trabeculae and form a framework for the lymphoid tissue and *lymphocytes* (white blood cells) in clusters called *lymphatic nodules.*

Inside the node is a *cortex* where most of the lymphocytes gather, and at the center is a *medulla,* which is less dense than the cortex but also contains lymphocytes. The outer cortex consists of lymphocytes arranged in masses called lymphatic nodules, which have central areas called *germinal centers* that produce the lymphocytes. Lymph fluid enters the node on its convex side through *afferent* (inbound) vessels that have valves opening only toward the node. Lymph circulates through the node, where it's filtered and then allowed to depart through *efferent* (outbound) vessels in the hilus with valves pointing exclusively away from the node. (If you have trouble remembering your afferent from your efferent, think of the "a" as standing for "access" and the "e" as standing for "exit".)

Although some lymph nodes are isolated from others, most nodes occur in groups, or clusters, particularly in the inguinal (groin), axillary (armpit), and mammary gland areas. The following are the primary lymph node regions:

    ✔ Head

    ✔ Neck

    ✔ Upper extremities

    ✔ Lower extremities

 Abdomen and pelvis

 Viscera

 Thorax

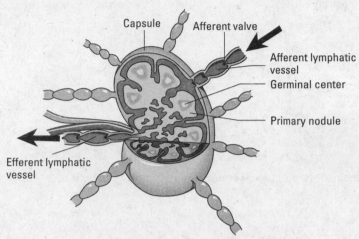

Figure 11-2:
A lymph
node.

Nodes in the head drain lymph from the scalp, upper neck, ear, parts of the eye, nose, and cheek. Lymph vessels from the head and the neck carry lymph to the nodes in the neck. Axillary nodes found in the armpit receive lymph from the upper arm, while the lymphatic vessels on the radial side of the arm supply nodes in the clavicle region. Lymph nodes in the chest region process lymph from the thoracic wall. Lymph nodes in the abdominal and pelvic region filter fluid from the lower body regions, reproductive organs, and thighs. Viscera nodes or gastric lymph nodes function in the drainage of the digestive organs. The inguinal nodes function to drain the lower extremities.

Each node acts like a filter bag filled with a network of thin, perforated sheets of tissue — a bit like cheesecloth — through which lymph must pass before moving on. White blood cells line the sheets of tissue, including several types that play critical roles in the body's immune defenses. This filtering action explains why, when infection first starts, lymph nodes often swell with the cellular activity of the immune system launching into battle with the invading microorganisms.

The cortex of each lymph node contains monocytes and two types of lymphocytes: *B cells* and *T cells.*

 Monocytes within the lymph nodes develop into large invader-eaters called *macrophages* that are capable of destroying a variety of microorganisms and sometimes even cancer cells.

 B cells don't attack pathogens directly but instead may produce molecules called *antibodies* that do the dirty work. Or they may instruct other cells called *phagocytes* (literally "cells that eat") to attack the invaders.

 T cells are lymphocytes that started out in the bone marrow but matured in the *thymus gland* (hence the name T cells) before moving on to the lymph nodes and spleen.

Think you have a node-tion (sorry!) about what's happening here? Test your knowledge:

Q. An area of the body where no lymph nodes are found is the

   a. Integument

   b. Liver

   c. Stomach

   d. Central nervous system (brain and spinal cord)

   e. Urinary system

A. The correct answer is the central nervous system (brain and spinal cord).

6. An encapsulated mass of lymph tissue connected to lymph vessels is a

   a. Tonsil

   b. Spleen

   c. Peyer's patch

   d. Thymus

   e. Lymph node

7. The lymph organ that has afferent and efferent lymph vessels is the

   a. Lymph node

   b. Tonsil

   c. Peyer's patches

   d. Spleen

   e. Thymus

8. Cells attached to the reticular fibers in a lymph node are

   a. Macrophages

   b. Monocytes

   c. Neutrophils

   d. Lymphocytes

   e. Basophils

9. A function of the lymph nodes is to

   a. Remove dead erythrocytes

   b. Produce bilirubin

   c. Produce lymphocytes

   d. Conserve iron

   e. Remove erythrocytes

10. The stroma of the lymph node does *not* consist of

   a. Capsule

   b. Hilus

   c. Cortex

   d. Trabeculae

   e. Villi

11. Lymphocytes are produced in the lymph nodules in the region called the

   a. Medulla

   b. Germinal center

   c. Lymphatic reservoir

   d. Trabeculae

   e. Hilus

12. The connective tissue fiber that forms the framework of the lymphoid tissue is

   a. Cartilaginous

   b. Collagenous

   c. Bone

   d. Elastic

   e. Reticular

13. When infection first starts, the lymph nodes tend to

   a. Recede

   b. Swell

   c. Multiply

   d. Divide

   e. Fade

14. T cells get their name because they start out in the bone marrow and mature in the

   a. Thigh

   b. Thyroid

   c. Thymus

   d. Tongue

   e. Tailbone

Having a Spleen-ded Time with the Lymphatic Organs

While the lymph nodes are the most numerous lymphatic organs, several other vital organs exist in the lymphatic system, including the spleen, thymus gland, and tonsils.

The spleen

The spleen, the largest lymphatic organ in the body, is a 5-inch, roughly egg-shaped structure to the left of and slightly behind the stomach. Like lymph nodes, it has a hilus through which the splenic artery, splenic vein, and efferent (remember "e" for "exit") vessels pass. Also like lymph nodes, the spleen's surrounded by a fibrous capsule that folds inward to section it off. Arterioles leading into each section are surrounded by masses of developing lymphocytes that give those areas of so-called *white pulp* their appearance. On the outer edges of each compartment, tissue called *red pulp* consists of blood-filled cavities. Unlike lymph nodes, the spleen doesn't have any afferent (access) lymph vessels, which means that it doesn't filter lymph, only blood.

Blood flows slowly through the spleen to allow it to remove microorganisms, exhausted erythrocytes (red blood cells), and any foreign material that may be in the stream. Among its various functions, the spleen can be a blood reservoir. When blood circulation drops while the body is at rest, the spleen's vessels can dilate to store any excess volume. Later, during exercise or if oxygen concentrations in the blood begin to drop, the spleen's blood vessels constrict and push any stored blood back into circulation.

But the spleen's primary role is as a biological recycling unit, capturing and breaking down defective and aged blood cells to reuse their components later. Iron stored by the spleen's macrophages goes to the bone marrow where it's turned into hemoglobin in new blood cells. By the same token, bilirubin for the liver is generated during breakdown of hemoglobin. The spleen produces red blood cells during embryonic development but shuts down that process after birth; in cases of severe anemia, the spleen sometimes starts up production of red blood cells again.

Fortunately, the spleen isn't considered a vital organ; if it's damaged or has to be surgically removed, the liver and bone marrow can pick up where the spleen leaves off.

T cell central: The thymus gland

Tucked just behind the breastbone and between the lungs in the upper chest, the thymus gland was a medical mystery until recent decades. Its two oblong lobes are largest at puberty when they weigh around 40 grams (somewhat less than an adult mouse). Through a process called *involution,* however, the gland atrophies and shrinks to roughly 6 grams by the time an adult is 65. (You can remember that term as the inverse of evolution.)

The thymus gland serves its most critical role — as a nursery for immature T lymphocytes, or T cells — during fetal development and the first few years of a human's life. Prior to birth, fetal bone marrow produces *lymphoblasts* (early stage lymphocytes) that migrate to the thymus. Shortly after birth and continuing until adolescence, the thymus secretes several hormones, collectively called *thymosin,* that prompt the early cells to mature into full-grown T cells that are immunocompetent, ready to go forth and conquer invading microorganisms. (These hormones are the reason the thymus is considered part of the endocrine system, too.)

As with other lymphatic structures, the thymus is surrounded by a fibrous capsule that dips inside to create chambers called *lobules.* Within each lobule is a cortex made of T cells held in place by reticular fibers and a central medulla of unusually onion-like layered epithelial cells called *thymic corpuscles,* or *Hassall's corpuscles,* as well as scattered lymphocytes.

Opening wide and moving along: The tonsils and Peyer's patches

Like the thymus gland, the tonsils, which are misunderstood masses of lymphoid tissue, are largest around puberty and tend to atrophy as an adult ages. Unlike the thymus, however, the tonsils don't secrete hormones but do produce lymphocytes and antibodies to protect against microorganisms that are inhaled or eaten. Although only two are visible on either side of the pharynx, there are actually six tonsils: the two you can identify, which are called *palatine tonsils;* two more called *adenoids* or pharyngeal tonsils in the wall of the pharynx; and two in the posterior one-third of the tongue called *lingual tonsils. Invaginations* (ridges) in the tonsils form pockets called *crypts,* which trap bacteria and other foreign matter.

Peyer's patches, also called *aggregate glands* or *agminate glands,* are masses of lymph tissue just below the surface of the ileum, the lowest section of the small intestine. When harmful microorganisms get into the intestine, Peyer's patches can mobilize an army of B cells and macrophages to fight off infection.

You've absorbed a lot in this section. See how much of it is getting caught in your filters:

15. Cells that remove foreign matter from the lymph in the lymph nodes are the

   a. Endothelial cells

   b. Erythrocytes

   c. Neutrophils

   d. Lymphocytes

   e. Macrophages

16. Which of the following is not a lymphatic organ?

   a. Tonsil

   b. Thymus

   c. Liver

   d. Spleen

17. The lymphatic organ that stops growth during adolescence and atrophies with aging is the

   a. Thymus

   b. Lymph nodes

   c. Tonsil

   d. Adenoids

   e. Spleen

18. Lymphoid tissue located in the pharynx that protects against inhaled or ingested pathogens and foreign substances is called the

   a. Thymus

   b. Tonsils

   c. Peyer's patches

   d. Spleen

   e. Lymph nodes

19. Lymphatic nodules found in the ileum of the small intestines are

    a. Tonsils

    b. Lymph nodes

    c. Thymus

    d. Macrophages

    e. Peyer's patches

20. The spleen is in close relation with the

    a. Stomach

    b. Liver

    c. Colon

    d. Kidney

    e. Duodenum

21. The lymphatic organ found in the superior mediastinum is the

    a. Tonsil

    b. Spleen

    c. Thymus

    d. Reticular formation

    e. Germinal center

22. Lymphocytes are the predominant cells in the

    a. Bone tissue

    b. Cartilage

    c. White pulp of the spleen

    d. Red pulp of the spleen

    e. Elastic connective tissue

23. The lymphatic organ responsible for removal of aged and defective blood cells is the

    a. Tonsil

    b. Spleen

    c. Peyer's patches

    d. Lymph nodes

    e. Thymus

24. The lymphatic organ that secretes hormones to make T lymphocytes immunocompetent is the

    a. Lymph node

    b. Tonsil

    c. Peyer's patch

    d. Spleen

    e. Thymus

25. Which of the following is not a true tonsil?

    a. Pharyngeal tonsil

    b. Palatine tonsils

    c. Adenoids

    d. Sublingual tonsil

    e. Lingual tonsil

26.–35. Mark the statement with a T if it's true or an F if it's false:

    26. _____ The lymph system offers an alternative route for the return of the tissue fluid to the bloodstream.

    27. _____ The fluid surrounding the cells that will enter the lymph capillaries is called interstitial fluid.

    28. _____ Lymph from the lymph vessels flows into the right thoracic duct and the left thoracic duct.

    29. _____ The lymph capillaries are composed of loosely overlapping reticular fiber.

    30. _____ The spleen filters both lymph and blood.

    31. _____ The thymus gland is functional in the early years of life and is most active in old age.

    32. _____ Tonsils function to protect against pathogens and foreign substances that are inhaled or ingested.

    33. _____ The spleen functions in the removal of aged and defective blood cells and platelets from the blood.

    34. _____ The gastric lymph nodes drain the lymphatic vessels on the radial side of the arm.

    35. _____ The thoracic duct originates from a triangular sac called the chyle cistern (or cisterna chyli).

36.–41. Fill in the blanks:

    36. The bilobular thymus gland is located in the _____.

    37. Once in the lymphatic system, interstitial fluid becomes known as _____.

    38. _____ are masses of lymphatic nodules found in the distal portion of the small intestines.

    39. The largest lymphatic organ in the body is the _____.

    40. The lymph nodes of the lower extremities drain the _____, _____, and _____.

    41. In the center of the nodules of the lymph node are areas called _____.

Answers to Questions on the Lymphatic System

The following are answers to the practice questions presented in this chapter.

1   The lymphatic system plays an important role in regulating **b. interstitial fluid protein.** By keeping the interstitial fluid volume between tissue cells in balance, the lymphatic system also keeps the body in homeostasis.

2   Terminated vessels that return plasma proteins to the blood are **c. lymph capillaries.** "Terminated vessels" is a technical way of saying they're a one-way path straight back to the source.

3   The thoracic duct does *not* drain lymph from the **b. right side of the head.** In fact, that's one of the very few areas the thoracic duct doesn't drain.

4   The largest lymphatic vessel in the body is the **c. thoracic duct.** Yes, this duct is the largest lymphatic *vessel.* "Spleen" isn't the correct answer because that's the largest lymphatic *organ.*

5   The lymphatic system does *not* function to **d. produce erythrocytes.** Those are red blood cells, which develop in the bone marrow.

6   An encapsulated mass of lymph tissue connected to lymph vessels is a **e. lymph node.** Don't let a $5 word like "encapsulated" fool you. It just means that the mass is wrapped in connective tissue.

7   The lymph organ that has afferent and efferent lymph vessels is the **a. lymph node.** The thymus and spleen have no inbound (afferent) vessels, and Peyer's patches and tonsils don't have much to do with lymph circulation.

8   Cells attached to the reticular fibers in a lymph node are **d. lymphocytes.** The reticular fibers create a net on which these cells can cluster.

9   A function of the lymph nodes is to **c. produce lymphocytes.** That's one of their two primary functions.

10   The stroma of the lymph node does *not* consist of **e. villi.** Nope, no finger-like projections here.

11   Lymphocytes are produced in the lymph nodules in the region called the **b. germinal center.** That's the heart of lymphocyte production in a nodule.

12   The connective tissue fiber that forms the framework of the lymphoid tissue is **e. reticular.** It provides both a tissue framework and a type of netting to hold clusters of lymphocytes.

13   When infection first starts, the lymph nodes tend to **b. swell.** This reaction occurs as the battle begins in your immune system at the cellular level.

14   T cells get their name because they start out in the bone marrow and mature in the **c. thymus.**

15   Cells that remove foreign matter from the lymph in the lymph nodes are the **e. macrophages.** These are the mature monocytes that can engulf a microorganism.

16   Which of the following is not a lymphatic organ? **c. Liver.** No lymph fluid here.

17 The lymphatic organ that stops growth during adolescence and atrophies with aging is the **a. thymus.**

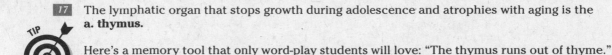

Here's a memory tool that only word-play students will love: "The thymus runs out of thyme."

18 Lymphoid tissue located in the pharynx that protects against inhaled or ingested pathogens and foreign substances is called the **b. tonsils.** When you remember where the pharynx is — the back of the throat — this question becomes more obvious.

19 Lymphatic nodules found in the ileum of the small intestines are **e. Peyer's patches.** It's almost like they're "patched" onto the ileum.

20 The spleen is in close relation with the **a. stomach.** It's certainly nearest to the stomach.

21 The lymphatic organ found in the superior mediastinum is the **c. thymus.**

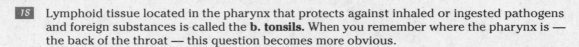

Break this question into parts and it becomes easier to locate which gland is being referenced: *Superior* means "upper," *media–* means "middle" (or "midline"), and *–stinum* refers to the sternum, or breastbone.

22 Lymphocytes are the predominant cells in the **c. white pulp of the spleen.** They're what give it its whitish color.

23 The lymphatic organ responsible for removal of aged and defective blood cells is the **b. spleen.** It recycles critical components from the spent blood cells and sends them to the bone marrow to be turned into fresh cells.

24 The lymphatic organ that secretes hormones to make T lymphocytes immunocompetent is the **e. thymus.** It's where these cells get the "T" in their name.

25 Which of the following is not a true tonsil? **d. Sublingual tonsil.** If it's not pharyngeal, palantine, or lingual, it's not a real tonsil.

26 The lymph system offers an alternative route for the return of the tissue fluid to the bloodstream. **True**

27 The fluid surrounding the cells that will enter the lymph capillaries is called interstitial fluid. **True**

28 Lymph from the lymph vessels flows into the right thoracic duct and the left thoracic duct. **False.** There is no right thoracic duct, only a right lymphatic duct.

29 The lymph capillaries are composed of loosely overlapping reticular fiber. **False.** The lymph capillaries actually are composed of overlapping endothelial cells.

30 The spleen filters both lymph and blood. **False**

31 The thymus gland is functional in the early years of life and is most active in old age. **False.** The opposite is true; the thymus is practically nonexistent in old age.

32 Tonsils function to protect against pathogens and foreign substances that are inhaled or ingested. **True**

33 The spleen functions in the removal of aged and defective blood cells and platelets from the blood. **True**

34 The gastric lymph nodes drain the lymphatic vessels on the radial side of the arm. **False.** This statement doesn't make much sense because "gastric" refers to the digestive system.

35 The thoracic duct originates from a triangular sac called the chyle cistern (or cisterna chyli). **True**

36 The bilobular thymus gland is located in the **superior mediastinum.**

37 Once in the lymphatic system, interstitial fluid becomes known as **lymph.** What else could it be called?

38 **Peyer's patches** are masses of lymphatic nodules found in the distal portion of the small intestines. Don't let that "distal" fool you; just think of it as "distant."

39 The largest lymphatic organ in the body is the **spleen.** You may be tempted to write "thoracic duct" here, but that's incorrect because the duct is the largest vessel, not the largest organ.

40 The lymph nodes of the lower extremities drain the **knee, leg, and foot.** It's a dead giveaway seeing as how those are all part of the lower extremities.

41 In the center of the nodules of the lymph node are areas called **germinal centers.** When you read "germinal," think of the word "germinate," and then think of a place where lymphocytes can sprout and mature.

Chapter 12

Filtering Out the Junk: The Urinary System

. . . .

. . . .

*I*f you read Chapter 9 on the digestive system, you may be chewing on the idea that undigested food is the body's primary waste product. But it's not — that title belongs to urine. We make more of it than we do feces — in fact, our bodies are making small amounts of urine all the time — and we release it more often throughout the day. Most important, urine captures all the leftovers from our cells' metabolic activities and jettisons them before they can build up and become toxic. In addition, urine helps maintain *homeostasis,* or the proper balance of body fluids.

In short, the urinary system

    ✔ Excretes useless and harmful material that it filters from blood plasma, including urea, uric acid, creatinine, and various salts

    ✔ Removes excess materials, particularly anything normally present in the blood that builds up to excessive levels

    ✔ Maintains proper osmotic pressure, or fluid balance, by eliminating excess water when concentration rises too high at the tissue level

In this chapter, we look at how the urinary system collects, manages, and excretes the waste that the body's cells produce as they go about busily metabolizing all day. You practice identifying parts of the kidneys, ureter, urinary bladder, and urethra.

Examining the Kidneys, the Body's Filters

The kidneys are nonstop filters that sift through 1.2 liters of blood per minute. Humans have a pair of kidneys just above the waist (lumbar region) toward the back of the abdominal cavity. While sometimes the same size, the left kidney tends to be a bit larger than the right. The last two pairs of ribs surround and protect each kidney, and a layer of fat, called *perirenal fat,* provides additional cushioning. Kidneys are *retroperitoneal,* which means that they're posterior to the peritoneum. The *renal capsule,* or outer lining of the kidney, is a layer of collagen fibers; these fibers extend outward to anchor the organ to surrounding structures.

Each kidney is dark red, about 4½ inches long, and shaped like a bean (hence the type of legumes called kidney beans). The portion of the bean that folds in on itself, referred to as the *medial border,* is concave with a deep depression in it called the *hilus,* or *hilum.* The hilus opens into a fat-filled space called the *renal sinus,* which in turn contains the *renal pelvis, renal calices,* blood vessels, nerves, and fat. The *renal artery* and *renal vein,* which provide the kidney's blood supply, as well as the ureter that carries urine to the bladder leave the kidney through the hilus.

Immediately below the renal capsule is a granular layer called the *renal cortex,* and just below that is an inner layer called the *medulla* that folds into anywhere from 8 to 18 conical projections called the *renal pyramids.* Between the pyramids are *renal columns* that extend from the cortex inward to the renal sinus. The tips of these pyramids, the *renal papillae,* empty their contents into a collecting area called the *minor calyx.* It's one of several sac-like structures referred to as the *minor* and *major calyces* which form the start of the urinary tract's "plumbing" system and collect urine transmitted through the papillae from the cortex and medulla. Although the number varies between individuals, generally each of two or three major calyces branches into four or five minor calyces, with a single minor calyx surrounding the papilla of one pyramid. Urine passes through the minor calyx into its major calyx and then on into the ureter for the trip to the bladder.

Going molecular

At the microscopic level, each kidney contains hundreds of thousands of tiny tubes known as *uriniferous tubules,* or *nephrons.* These are the primary functional units of the urinary system. At one end, each nephron is closed off and folded into a small double-cupped structure called a *Bowman's capsule,* or the *glomerular capsule,* where the actual process of filtration occurs. Leading away from the capsule, the nephron forms into the *first* or *proximal convoluted tubule* (PCT), which is lined with cuboidal epithelial cells having microvilli brush borders that increase the area of absorption. This tube straightens to form a structure called the *descending loop of Henle* and then bends back in a hairpin turn into another structure called the *ascending loop of Henle.* After that, the tube becomes convoluted again, forming the *second* or *distal convoluted tubule* (DCT), which is made of the same types of cells as the first tubule but without any microvilli. This tubule connects to a collecting tubule that it shares with the output ends of many other nephrons. The collecting tubules open into the minor calyces of the renal pelvis, which in turn open into the major calyces.

Because of their role as the body's key filters, the kidneys receive about 20 percent of all the blood pumped by the heart each minute. A large branch of the abdominal aorta, called the *renal artery,* carries that blood to them. After branching into smaller and smaller vessels, the blood eventually enters *afferent arterioles,* each of which branches into tufts of five to eight capillaries called a *glomerulus* (the plural is *glomeruli*) inside the Bowman's capsule. After picking up waste products from the filters inside the wall of the capsule, the capillaries come back together to form *efferent arterioles,* which then branch to form the *peritubular,* or *second, capillary bed* surrounding the convoluted tubules, the loop of Henle, and the collecting tubule. The capillaries come together once again to form a small vein that empties blood into the renal vein to depart the kidneys.

Each glomerulus and its surrounding Bowman's capsule make up a single *renal corpuscle* where basic filtration takes place. Like all capillaries, glomeruli have thin, membranous walls, but unlike their capillary cousins elsewhere, these vessels have unusually large pores called *fenestrations* or *fenestrae* (from the Latin word *fenestra* for "window").

Focusing on filtering

To understand how the renal corpuscles work, think of an espresso machine: Water is forced under pressure through a sieve containing ground coffee beans, and a filtrate called brewed coffee trickles out the other end. Something similar takes place in the renal corpuscles. Hydrostatic pressure forces fluids across the glomerular membranes, which capture about 125 milliliters of material per minute in the Bowman's capsules.

Selective reabsorption occurs (mostly in the PCT) as these filtered materials then move through the nephrons' network of tubules, returning the bulk of the water and much of the needed materials back to the bloodstream through the peritubular capillary bed surrounding the nephrons' structures.

So despite 125 milliliters of material coming out of the blood every minute, only 1 milliliter of urine is generated each minute. This is a matter of simple subtraction: Reabsorption of about 100 milliliters per minute takes place in the proximal convoluted tubules. The loop of Henle returns 7 milliliters per minute more. The distal convoluted tubules return 12 milliliters, and the collecting tubules return about 5 milliliters. Voila! That totals 124 milliliters of reabsorption per minute and explains the 1 milliliter of urine that comes out when all is said and done.

While all this filtering and absorption is going on, the kidneys also sometimes secrete an enzyme called *renin* (also known by its more complicated chemical name of *angiotensinogenase*) that converts a peptide generated in the liver, called *angiotensinogen,* into *angiotensin I.* Angiotensin I moves into the lungs where a converting enzyme turns it into *angiotensin II,* a potent *vasoconstrictor.* Say what? Try this explanation, instead: The kidneys work to ensure that systemic blood pressure remains high enough for them to do their filtering job properly. That's what a vasoconstrictor is: a substance that causes blood vessels to narrow, increasing the pressure of the fluids moving through them. Rising blood pressure also triggers the adrenal glands perched atop each kidney to release *aldosterone,* causing the renal tubules to absorb more sodium and pumping up blood volume. The pituitary gland also plays a role in urine production by releasing an antidiuretic hormone (ADH) that causes water retention at the kidneys and elevated blood pressure.

You've absorbed a lot in the last few paragraphs. See how much of it is getting caught in your filters:

1.–5. Match the anatomical terms with their descriptions.

1. _____ Cortex

2. _____ Medulla

3. _____ Renal pelvis

4. _____ Calyx

5. _____ Collecting tubule

a. Composed of folds forming the renal pyramids

b. Granular outer layer

c. Irregular sac-like structures for collecting urine in the renal pelvis

d. Transports urine from the cortex

e. Found in the renal sinus

6. The functional kidney is responsible for

a. Removal of carbon dioxide

b. Excretion of waste

c. Removal of excessive materials in the blood

d. All of the above

e. Only b and c

7. The ureter leaves the kidney through the depression known as the

a. Hilus

b. Pyramids

c. Cortex

d. Medulla

e. Calyx

8. The human kidney is not included in the abdominal cavity, which makes it

   a. Retroperitoneal

   b. Parietal

   c. Endocoelomic

   d. Exterocoelomic

   e. None of these

9. The functional *unit* of the kidney is the

   a. Glomerulus

   b. Henle's loop

   c. Collecting tubule

   d. Nephron

10. The correct sequence for removal of material from the blood through the nephron is

   a. Afferent arteriole → Glomerulus → Proximal convoluted tubule → Loop of Henle → Distal convoluted tubule → Collecting tubule

   b. Afferent arteriole → Glomerulus → Distal convoluted tubule → Loop of Henle → Proximal convoluted tubule → Collecting tubule

   c. Afferent arteriole → Collecting tubule → Glomerulus → Proximal convoluted tubule → Loop of Henle → Distal convoluted tubule

   d. Efferent arteriole → Proximal convoluted tubule → Glomerulus → Loop of Henle → Distal convoluted tubule → Collecting tubule

11. Brush borders of villi are found primarily in the

   a. Ascending loop of Henle

   b. Proximal convoluted tubule

   c. Bowman's capsule

   d. Descending loop of Henle

   e. Collecting tubules

12. Filtration occurs primarily in the

   a. Distal convoluted tubules

   b. Loop of Henle

   c. Collecting tubule

   d. Glomerulus

   e. Proximal convoluted tubules

13. Reabsorption occurs primarily in the

   a. Collecting tubules

   b. Ureter

   c. Glomerulus

   d. Proximal convoluted tubules

   e. Distal convoluted tubules

Getting Rid of the Waste

After your kidneys filter out the junk, it's time to deliver it to the bladder. Here is how that's done.

Surfing the ureters

Ureters are narrow, muscular tubes through which the collected waste travels. About 10 inches long, each ureter descends from a kidney to the posterior lower third of the bladder. Like the kidneys themselves, the ureters are behind the peritoneum outside the abdominal cavity, so the term *retroperitoneal* applies to them, too. The inner wall of the ureter is a simple mucous membrane. It also has a middle layer of smooth muscle tissue that propels the urine by peristalsis — the same process that moves food through the digestive system. So rather than trickling into the bladder, urine arrives in small spurts as the muscular contractions force it down. The tube is surrounded by an outer fibrous layer of connective tissue that supports it during peristalsis.

Ballooning the bladder

The urinary *bladder* is a large muscular bag that lies in the pelvis behind the pubis bones. In female humans, it's beneath the uterus and in front of the vagina. In male humans, the bladder lies between the rectum and the symphysis pubis. There are three openings in the bladder: two on the back side where the ureters enter and one on the front for the *urethra,* the tube that carries urine outside the body. The bladder's *trigone* is the triangular area between these three openings. The *neck* of the bladder surrounds the urethral attachment, and the *internal sphincter* (smooth muscle that provides involuntary control) encircles the junction between the urethra and the bladder.

Inside, the bladder is lined with highly elastic transitional epithelium tissue. When full, the bladder's lining is smooth and stretched; when empty, the lining lies in a series of folds called *rugae* (just as the stomach does). When the bladder fills, the increased pressure stimulates the organ's stretch receptors, prompting the individual to urinate.

The male and female urethras

Both males and females have a *urethra,* the tube that carries urine from the bladder to a body opening, or orifice. Both males and females have an internal sphincter controlled by the autonomic nervous system and composed of smooth muscle to guard the exit from the bladder. Both males and females also have an external sphincter composed of circular striated muscle that's under voluntary control. But as we all well know, the exterior plumbing is rather different.

The female urethra is about one and a half inches long and lies close to the vagina's anterior (front) wall. It opens just in front of the vaginal opening. The external sphincter for the female urethra lies just inside the urethra's exit point.

The male urethra is about 8 inches long and carries a different name as it passes through each of three regions:

> ✔ The *prostatic urethra* leaving the bladder contains the internal sphincter and passes through the prostate gland. Several openings appear in this region of the urethra, including a small opening where sperm from the vas deferens and ejaculatory duct enters, and prostatic ducts where fluid from the prostate enters.
>
> ✔ The *membranous urethra* is a small 1- or 2-centimeter portion that contains the external sphincter and penetrates the pelvic floor. The tiny *bulbourethral (or Cowper's) glands* lie on either side of this region.
>
> ✔ The *cavernous urethra*, also known as the *spongy urethra*, runs the length of the penis on its ventral surface through the corpus spongiosum, ending at a vertical slit at the end of the penis. Ducts from the Cowper's glands enter at this region.

Now test your knowledge of how the human body gets rid of its waste:

0. The separation of the reproductive and urinary systems is complete in the human

   a. Male

   b. Female

   c. Both

A. The correct answer is female. The male urethra runs through the same "plumbing" as the male reproductive system.

14. The interior of the bladder is lined with elastic

   a. Ciliated columnar epithelium

   b. Transitional epithelium

   c. Cuboidal epithelium

   d. White fibrous connective tissue

   e. Squamous epithelium

15. The urine formed in the kidney is passed down the ureter by

   a. Fibrillation

   b. Flexure

   c. Gravity

   d. Gestation

   e. Peristaltic contractions

16. The circular striated muscle known as the external sphincter is found in the

   a. Kidney

   b. Ureter

   c. Uterus

   d. Urethra

   e. Bladder

17. The internal sphincter found at the junction of the bladder neck and the urethra is composed of

   a. Striated muscle tissue

   b. Squamous endothelial tissue

   c. Mucous membrane

   d. Transitional epithelium

   e. Smooth muscle tissue

Spelling Relief: Urination

Urination, known by the medical term *micturition,* occurs when the bladder is emptied through the urethra. Although urine is created continuously, it's stored in the bladder until the individual finds a convenient time to release it. Mucus produced in the bladder's lining protects its walls from any acidic or alkaline effects of the stored urine. When there is about 200 milliliters of urine distending the bladder walls, stretch receptors transmit impulses to warn that the bladder is filling. *Afferent* impulses are transmitted to the spinal cord, and *efferent* impulses return to the bladder, forming a reflex arc that causes the internal sphincter to relax and the muscular layer of the bladder to contract, forcing urine into the urethra. The afferent impulses continue up the spinal cord to the brain, creating the urge to urinate. Because the external sphincter is composed of skeletal muscle tissue, no urine usually is released until the individual voluntarily opens the sphincter.

18.–23. Use the terms that follow to identify the parts of the kidney in Figure 12-1.

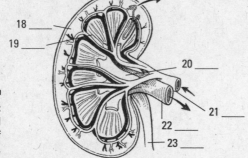

Figure 12-1: Internal anatomy of the kidney.

LifeART Image Copyright 2007 ©. Wolters Kluwer Health — Lippincott Williams & Wilkins

   a. Renal vein

   b. Cortex

   c. Ureter

   d. Renal pelvis

   e. Renal artery

   f. Medulla (renal pyramids)

24.–31. Use the terms that follow to identify the nephron structures in Figure 12-2.

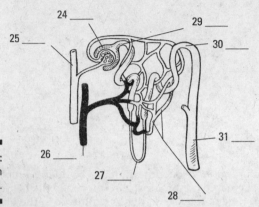

24 _____
25 _____
29 _____
30 _____
26 _____
27 _____
28 _____
31 _____

Figure 12-2:
Nephron
structures.

LifeART Image Copyright 2007 ©. Wolters Kluwer Health — Lippincott Williams & Wilkins

a. Loop of Henle

b. Glomerular (Bowman's) capsule

c. Collecting tubule

d. Proximal convoluted tubule

e. Renal vein

f. Second capillary bed

g. Distal convoluted tubule

h. Renal artery

Answers to Questions on the Urinary System

The following are answers to the practice questions presented in this chapter.

1. Cortex: **b. Granular outer layer**

2. Medulla: **a. Composed of folds forming the renal pyramids**

3. Renal pelvis: **e. Found in the renal sinus**

4. Calyx: **c. Irregular sac-like structures for collecting urine in the renal pelvis**

5. Collecting tubule: **d. Transports urine from the cortex**

6. The functional kidney is responsible for **e. only b and c (excretion of waste and removal of excessive materials in the blood).** The other answer option can't be correct because carbon dioxide exits the body through the lungs.

7. The ureter leaves the kidney through the depression known as the **a. hilus.** That's the part where the "bean" folds in on itself.

8. The human kidney is not included in the abdominal cavity, which makes it **a. retroperitoneal.** *Peritoneal* refers to the peritoneum, the membrane lining the abdominal cavity; and *retro* can be defined as "situated behind."

9. The functional *unit* of the kidney is the **d. nephron.** Each nephron contains a series of the parts needed to do the kidney's filtering job.

10. The correct sequence for removal of material from the blood through the nephron is **a. Afferent arteriole → Glomerulus → Proximal convoluted tubule → Loop of Henle → Distal convoluted tubule → Collecting tubule.** In short, blood comes through the artery (arteriole) and material gloms onto the nephron before twisting through the near (proximal) tubes, looping the loop, twisting through the distant (distal) tubes, and collecting itself at the other end. Try remembering artery-glom-proxy-loop-distant-collect.

11. Brush borders of villi are found primarily in the **b. proximal convoluted tubule.** Those brush borders provide extra surface area for reabsorption, so it makes sense that they congregate in the first area after filtration.

12. Filtration occurs primarily in the **d. glomerulus.** The glomerulus is a collection of capillaries with big pores, so think of it as the initial filtering sieve.

13. Reabsorption occurs primarily in the **d. proximal convoluted tubules.** These tubules have the most surface area with all those villi brush borders, so they reabsorb the most.

14. The interior of the bladder is lined with elastic **b. transitional epithelium.** It's transitional because it needs to be able to stretch and collapse as needed.

15. The urine formed in the kidney is passed down the ureter by **e. peristaltic contractions.** It's the same action that moves food through the digestive system.

16. The circular striated muscle known as the external sphincter is found in the **d. urethra.** If it's external, it has to be in the tube that heads outside.

17 The internal sphincter found at the junction of the bladder neck and the urethra is composed of **e. smooth muscle tissue.** The internal sphincter is smooth muscle tissue that prevents urine leakage from the bladder.

18 – 23 Following is how Figure 12-1, the internal anatomy of the kidney, should be labeled.

> 18. **b. Cortex**; 19. **f. Medulla (renal pyramids)**; 20. **d. Renal pelvis**; 21. **e. Renal artery**; 22. **a. Renal vein**; 23. **c. Ureter**

24 – 31 Following is how Figure 12-2, the nephron, should be labeled.

> 24. **b. Glomerular (Bowman's) capsule**; 25. **h. Renal artery**; 26. **e. Renal vein**; 27. **a. Loop of Henle**; 28. **f. Secondary capillary bed**; 29. **d. Proximal convoluted tubule**; 30. **g. Distal convoluted tubule**; 31. **c. Collecting tubule**

Part IV
Survival of the Species

The 5th Wave   By Rich Tennant

"I never realized trying to have a baby would mean replacing the soft music and candle light with an ovulation strip, a thermometer, and a starter pistol."

In this part . . .

In Part I, we review how cells perpetuate themselves. So in this part, we look at how the entire human species ensures its future through that process of cell division. We go with the guys first because — let's face it — their biological role in reproduction isn't as involved as women's. Then we explain reproduction on the female side of the equation, including a review of the human life cycle from birth to death.

Chapter 13

Why Ask Y?: The Male Reproductive System

In This Chapter

▶ Explaining the parts of male reproduction

▶ Understanding meiosis and what happens to chromosomes

*I*ndividually, humans don't need to reproduce to survive. But to survive as a species, a number of individuals must produce and nurture a next generation, carrying their uniqueness forward in the genetic pool. Humans are born with the necessary organs to do just that.

In this chapter, you get an overview of the parts and functions of the male reproductive system, along with plenty of practice questions to test your knowledge. (We cover the guys first because their role in the basic reproduction equation isn't nearly as long or complex as it is for their mates. We address the female reproductive system in Chapter 14.)

Identifying the Parts of the Male Reproductive System

On the outside, the male reproductive parts, which you can see in Figure 13-1, are straight-forward — a *penis* and a *scrotum*. At birth, the apex of the penis is enclosed in a fold of skin called the *prepuce,* or *foreskin,* which often is removed during a surgery called *circumcision*.

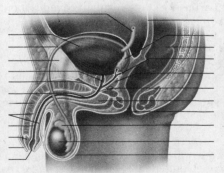

Figure 13-1: The male reproductive system.

Ureter — Peritoneum
Urinary bladder — Seminal vesicle
Ductus deferens — Rectum
Pubis — Ejaculatory duct
Prostatic urethra — Prostate gland
Urogenital diaphragm — Bulbourethral gland
Membranus urethra — Bulb of penis
Corpus cavernosum — Anus
Corpus spongiosum — Epididymis
Cavernous urethra — Testis
Glans penis — Scrotum
Prepuce
External urethral orifice

Illustration by Imagineering Media Services Inc.

The scrotum is a pouch of skin divided in half on the surface by a ridge called a *raphe* that continues up along the underside of the penis and down all the way to the anus. The left side of the scrotum tends to hang lower than the right side to accommodate a longer *spermatic cord,* which we explain later in this section.

There are two scrotal layers: the *integument,* or outer skin layer, and the *dartos tunic,* an inner smooth muscle layer that contracts when cold and elongates when warm. Why? That has to do with the two *testes* (the singular is *testis*) inside (see Figure 13-2). These small ovoid glands, also referred to as *testicles,* need to be a bit cooler than body temperature in order to produce viable sperm for reproduction. When the dartos tunic becomes cold, such as when a man is swimming, it contracts and draws the testes toward the body for warmth. When the dartos tunic becomes overly warm, it slackens to allow the testes to hang farther away from the heat of the body.

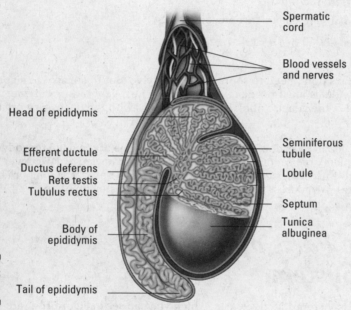

Figure 13-2:
Testis.

Illustration by Imagineering Media Services Inc.

A fibrous capsule called the *tunica albuginea* encases each testis and extends into the gland forming incomplete *septa* (partitions), which divide the testis into about 200 *lobules.* These compartments contain small, coiled *seminiferous tubules* where sperm are produced by *spermatogenesis,* or *meiosis,* which we review in the next section.

Distributed in gaps between the tubules are interstitial cells called *Leydig cells* that produce the male sex hormone *testosterone.* The tubules of each lobule come together in an area called the *mediastinum testis* and straighten into *tubuli recti* before forming a network called the *rete testis* that leads to the *efferent ducts* (also called *ductules*). These ducts carry sperm to an extremely long (about 20 feet), tightly coiled tube called the *epididymis* for storage.

The epididymis merges with the *ductus deferens,* or *vas deferens,* which carries sperm up into the spermatic cord, which also encases the testicular artery and vein, lymphatic vessels, and nerves. Convoluted pouches called *seminal vesicles* lie behind the base of the bladder and secrete an alkaline fluid containing fructose, vitamins, amino acids, and prostaglandins to nourish sperm as it enters the *ejaculatory duct.*

From there the mixture containing sperm enters the *prostatic urethra* that's surrounded by the *prostate gland.* This gland secretes a thin, opalescent substance that precedes the sperm in an ejaculation. The alkaline nature of this substance reduces the natural acidity of the female's vagina to prepare it to receive the sperm.

Two yellowish pea-sized bodies called *Cowper's glands,* or *bulbourethral glands,* lie on either side of the urethra and secrete a clear alkaline lubricant prior to ejaculation; it neutralizes the acidity of the urethra and acts as a lubricant for the penis. Once all the glands have added protective and nourishing fluids to the 400 to 500 million departing sperm, the mixture is known as *seminal fluid* or *semen.*

During sexual arousal, two columns of spongy erectile tissue in the penis — the *corpus spongiosum penis* and the *corpus cavernosum penis* — swell with blood to make it rigid and capable of entering the female's vagina. At the time of ejaculation, smooth muscles in the wall of the epididymis force sperm through the ductus deferens, located in the *inguinal canal,* toward the urinary bladder. After mixing with the secretions from the seminal vesicles and the prostate gland, the semen travels along the urethra and out a vertical slit in the *glans penis,* or head of the penis.

See how familiar you are with the male anatomy by tackling these practice questions:

1. The cutaneous pouch containing the testes and part of the spermatic cord is the

    a. Scrotum

    b. Ejaculatory pouch

    c. Ductus deferens

    d. Seminal vesicle

    e. Epididymis

2. The scrotum adjusts to surrounding temperatures through the action of the

    a. Testes

    b. Bulbourethral glands

    c. Dartos tunic

    d. Prostatic urethra

3. Spermatogenesis occurs in the

    a. Inguinal cells

    b. Interstitial cells

    c. Seminiferous tubules

    d. Rete testis

    e. Epididymis

4. Testosterone is produced in the

    a. Seminiferous tubules

    b. Epididymis

    c. Adenohypophysis

    d. Subtentacular cells

    e. Leydig cells

5. Select the correct sequence for the movement of sperm:

a. Seminiferous tubules → Tubuli recti → Rete testis → Epididymis → Ductus deferens → Ejaculatory duct → Urethra

b. Seminiferous tubules → Tubuli recti → Rete testis → Ejaculatory duct → Epididymis → Duct deferens → Urethra

c. Epididymis → Ejaculatory duct → Tubuli recti → Rete testis → Seminiferous tubules → Ductus deferens → Urethra

d. Seminiferous tubules → Ejaculatory duct → Tubuli recti → Rete testis → Epididymis → Ductus deferens → Urethra

6. Which of the following does *not* add a secretion to the sperm as it moves through the reproductive ducts?

a. Interstitial cells

b. Epididymis

c. Seminal vesicle

d. Prostate

7. The convoluted tube that stores sperm is called the

a. Seminiferous tubule

b. Rete testis

c. Spermatic cord

d. Epididymis

e. Ductus deferens

8. The fluid accompanying the sperm is called the

a. Stroma

b. Semen

c. Prepuce

d. Inguinal

e. Perineum

9. An average ejaculation will contain sperm numbering approximately

a. 40 to 50 million

b. 400 to 500 million

c. 400 to 500

d. 4 to 5 million

10. A thin, milky liquid imparting alkaline characteristics to the seminal fluid is produced by the

a. Seminal vesicle

b. Interstitial cells

c. Sertoli cells

d. Bulbourethral gland

e. Prostate gland

Packaging the Chromosomes for Delivery

Sperm, the male sex cell, is produced during a process called *meiosis* (which also produces the female sex cell, or *ovum*). Meiosis involves two divisions.

✔ The first, a *reduction division,* divides a single *diploid* cell with two sets of chromosomes into two *haploid* cells with only one set each.

✔ The second process is a division by *mitosis* that divides the two haploid cells into four cells with a single set of chromosomes each.

Review the reproduction terms in Table 13-1.

Table 13-1	Reproduction Terms to Know		
Terms That Vary By Gender	**General Term**	**Male Term**	**Female Term**
Sex organs	Gonads	Testes	Ovaries
Original cell	Gametocyte	Spermatocyte	Oocyte
Meiosis	Gametogenesis	Spermatogenesis	Oogenesis
Sex cell	Gamete	Spermatozoa (Sperm)	Ovum

A *diploid cell* (or 2N) has two sets of chromosomes, whereas a *haploid cell* (or 1N) has one set of chromosomes.

Meiosis, which you can see in Figure 13-3, is a continuous process. Once it starts, it doesn't stop until gametes are formed. Meiosis is described in a series of stages as follows (for more on the terminology of cells, see Chapter 2):

1. **Interphase:** The original diploid cell — called a *spermatocyte* in a man and an *oocyte* in a woman — is said to be in a "resting stage," but it actually undergoes constant activity. Just before it starts to divide, the DNA molecules in the *chromanemata* (chromatin network) duplicate.

2. **Prophase I:** Structures disappear from the nucleus, including the nuclear membrane, nucleoplasm, and nucleoli. The cell's centrosome divides into two *centrioles* that move to the ends of the nucleus and form poles. Structures begin to appear in the nuclear region, including *spindles* (protein filaments that extend between the poles) and *asters*, or *astral rays* (protein filaments that extend from the poles into the cytoplasm). The chromanemata contract, forming chromosomes. Those chromosomes then start to divide into two *chromatids* but remain attached by the centromere. *Homologous* chromosomes that contain the same genetic information pair up and go into *synapsis,* twisting around each other to form a *tetrad* of four chromatids. These tetrads begin to migrate toward the *equatorial plane* (an imaginary line between the poles).

3. **Metaphase I:** The tetrads align on the equatorial plane, attaching to the spindles by the centromere.

4. **Anaphase I:** Homologous chromosomes separate by moving along the spindles to opposite poles. In late anaphase, a slight furrowing is apparent in the cytoplasm, initiating *cytokinesis* (the division of the cytoplasm).

Meiosis:

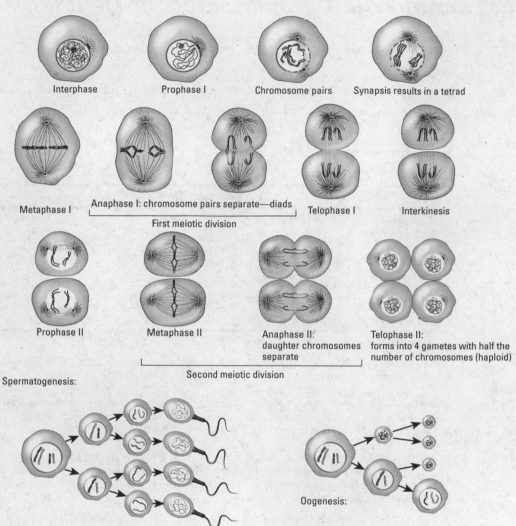

Interphase

Prophase I

Chromosome pairs

Synapsis results in a tetrad

Metaphase I

Anaphase I: chromosome pairs separate—diads

First meiotic division

Telophase I

Interkinesis

Prophase II

Metaphase II

Anaphase II: daughter chromosomes separate

Telophase II: forms into 4 gametes with half the number of chromosomes (haploid)

Second meiotic division

Spermatogenesis:

Oogenesis:

Figure 13-3: The stages of meiosis.

In meiosis in male (spermatogenesis) all four haploid cells become functional sperms.

In meiosis in female (oogenesis) only one of haploid cells becomes a functional egg.

Illustration by Imagineering Media Services Inc.

5. **Telophase I:** The contracted and divided homologous chromosomes are at opposite poles. Spindle and aster structures disappear, and a nuclear membrane and nucleoplasm begin to appear in each newly forming cell. Chromosomes remain as chromatids, still contracted and divided. The furrowing seen in anaphase continues to deepen, dividing the cytoplasm. In the male, the cytoplasm divides equally between the two cells. In the female, cytoplasmic division is unequal.

6. **Interkinesis:** The cytoplasm separates. Two genetically identical haploid cells are formed with half the number of chromosomes as the original cell. In the male, the cells are of equal size. In the female, one cell is large and the other is small.

7. **Prophase II:** The cells enter the second phase of meiosis. Once again, structures disappear from the nuclei and poles appear at the ends. Spindles and asters appear in the nuclear region. Chromosomes are already contracted and divided into chromatids attached by the centromere, and they begin to migrate toward the equatorial plane.

8. **Metaphase II:** The chromatids align on the equatorial plane and attach to the spindles by the centromere.

9. **Anaphase II:** The chromatids separate, becoming chromosomes that move along the spindles to the poles. A slight furrowing appears in the cytoplasm.

10. **Telophase II:** With the chromosomes at the poles, spindles and asters disappear while new nuclear structures appear. The chromosomes uncoil, returning to chromanemata and their chromatin network. Cytoplasmic division continues to deepen and each haploid cell divides, forming four cells.

At the end of this process, the male has four haploid sperm of equal size. As the sperm matures further, a *flagellum* (tail) develops. The female, on the other hand, has produced one large cell, the ovum, and three small cells called *polar bodies;* all four structures contain just one set of chromosomes. The polar bodies eventually disintegrate and the ovum becomes the functional cell. When fertilized by the sperm, the resulting *zygote* (fertilized egg) is diploid, containing two sets of chromosomes.

Think you've conquered this process? Find out by tackling these practice questions:

0. The metaphase II stage in meiosis involves

   a. The slipping of the centromere along the chromosome

   b. The alignment of the chromosomes on the equatorial plane

   c. The contraction of the chromosomes

   d. The disappearance of the nuclear membrane

   e. The polar migration of the chromosomes

A. The correct answer is the alignment of the chromosomes on the equatorial plane. Think "divide and conquer."

11. The process in sexual reproduction involving the union of gametes is called

   a. Fission

   b. Conjugation

   c. Invagination

   d. Fertilization

   e. Pollination

12. Gametes are formed by

   a. Interkinesis

   b. Cytokinesis

   c. Meiosis

   d. Mitosis

   e. Photosynthesis

13. A man has 46 chromosomes in a spermatocyte. How many chromosomes are in each sperm?

    a. 23 pairs

    b. 23

    c. 184

    d. 46

    e. 92

14. Synapsis, or side-by-side pairing, of homologous chromosomes

    a. Occurs in mitosis

    b. Completes fertilization

    c. Is followed immediately by the splitting of each centromere

    d. Signifies the end of prophase of the second meiotic division

    e. Occurs in meiosis

15. Anaphase I of meiosis is characterized by which of the following?

    a. Synapsed homologous chromosomes move toward the poles.

    b. DNA duplicates itself.

    c. Synapsis of homologous chromosomes occurs.

    d. Homologous chromosomes separate and move poleward with centromeres intact.

    e. Centromeres split, and chromosomes migrate toward the poles.

16. During ovum production, the three nonfunctional cells produced are called

    a. Diploid cells

    b. Male sex cells

    c. Polar bodies

    d. Somatic cells

    e. Cross-over gametes

17. The stage (or period) in meiosis between the first and second division is called

    a. Anaphase I

    b. Interphase

    c. Metaphase I

    d. Interkinesis

    e. Prophase II

18. After meiosis, each diploid spermatocyte has become

    a. Four haploid sperm of varying sizes

    b. One functional diploid sperm and three cells that eventually will disintegrate

    c. Four haploid sperm of equal size

    d. Two haploid sperm of equal size

    e. Three haploid flagellum and a diploid sperm

19. Complete the following worksheet on the stages of meiosis.

Draw the stages of meiosis: Describe the changes in each stage:

1.

Late Prophase I

2.

Metaphase I

3.

Anaphase I

4.

Teleophase I

5.

Interkinesis

6.

Prophase II

7.

Metaphase II

8.

Anaphase II

9.

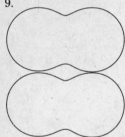

Teleophase II

10.

Sperm

11.

Ovum and Polar Bodies

Answers to Questions on the Male Reproductive System

The following are answers to the practice questions presented in this chapter.

1 The cutaneous pouch containing the testes and part of the spermatic cord is the **a. scrotum.** "Cutaneous" simply means "skin."

2 The scrotum adjusts to surrounding temperatures through the action of the **c. dartos tunic.** This is the inner smooth muscle layer of the scrotum.

3 Spermatogenesis occurs in the **c. seminiferous tubules.** Immature sperm cells line the walls of the tubules.

4 Testosterone is produced in the **e. Leydig cells.** "Interstitial cells" would also be correct because the word "interstitial" can be translated as "placed between." But alas, that's not one of the answer options.

5 Select the correct sequence for the movement of sperm: **a. Seminiferous tubules → Tubuli recti → Rete testis → Epididymis → Ductus deferens → Ejaculatory duct → Urethra.** The sperm develop in the coiled tubules, move through the straighter tubes (tubuli recti), continue across the network of the testis (rete testis) and into the epididymis (remember the really long tube), and travel past the ductus (or vas) deferens and the ejaculatory duct into the urethra.

6 Which of the following does *not* add a secretion to the sperm as it moves through the reproductive ducts? **a. Interstitial cells.** Interstitial cells secrete testosterone, which goes into the blood.

7 The convoluted tube that stores sperm is called the **d. epididymis.** The other answer options don't come into play until it's time to release semen.

8 The fluid accompanying the sperm is called the **b. semen.**

9 An average ejaculation will contain sperm numbering approximately **b. 400 to 500 million.** Keep in mind that sperm are microscopically small, so quite a few can fit in a tiny amount of semen.

10 A thin, milky liquid imparting alkaline characteristics to the seminal fluid is produced by the **e. prostate gland.**

11 The process in sexual reproduction involving the union of gametes is called **d. fertilization.** Oh, come now — fission? Pollination? Remember that we're talking human anatomy here!

12 Gametes are formed by **c. meiosis.** Gametes — sperm and ova — are the end goal of this process.

13 A man has 46 chromosomes in a spermatocyte. How many chromosomes are in each sperm? **b. 23.** Divide the number 46 in half.

14 Synapsis, or side-by-side pairing, of homologous chromosomes **e. occurs in meiosis.**

15 Anaphase I of meiosis is characterized by which of the following? **d. Homologous chromosomes separate and move poleward with centromeres intact.**

16 During ovum production, the three nonfunctional cells produced are called **c. polar bodies.** They eventually disintegrate.

17 The stage (or period) in meiosis between the first and second division is called **d. interkinesis.** *Inter–* means "between," and *–kinesis* means "motion," so it's clear that this phase is "between motions."

18 After meiosis, each diploid spermatocyte has become **c. four haploid sperm of equal size.** There's no such thing as a diploid sperm because as a sex cell, sperm carries only half the regular complement of 46 chromosomes. And because another division takes place after the initial division in meiosis, the final product of the process is four cells, not two.

19 Following is a summary of what should appear in your drawings and descriptions of the stages of meiosis. For further reference, check out Figure 13-3.

In the drawing for late prophase I, at least two pairs of homologous chromosomes should be shown grouped into tetrads (in truth, there are 23 pairs, but simplified illustrations tend to show just two). The description for prophase I should include reference to the tetrad formation. The drawing for metaphase I should show the equatorial plane (a center horizontal line) with the tetrads aligned along it. The illustration also should show spindles radiating from each pole, with the tetrads attached to them by their centromeres. The description should include reference to the equatorial plane, the poles, and the spindles.

The drawing for anaphase I should show the tetrads moving to the top and bottom of the cell along the spindles and the cytoplasm slowly beginning to divide. In telophase I, the division becomes more pronounced and two new nuclei form. As the process enters interkinesis, the cytoplasm pinches off into two cells.

During prophase II, which also is the start of the second meiotic division, the contracted and divided chromatids migrate toward a new equatorial plane. The drawing of metaphase II should show all chromatids aligned on the equatorial plane. For anaphase II, you should show the chromatids pulling apart into chromosomes and moving toward the poles. In the final stage, telophase II, you should draw new nuclei forming around the chromosomes.

Chapter 14

Carrying Life Forward: The Female Reproductive System

Men may have quite a few hard-working parts in their reproductive systems, but women are the ones truly responsible for survival of the species (biologically speaking, anyway). The female body prepares for reproduction every month for most of a woman's adult life, producing an ovum and then measuring out delicate levels of hormones to prepare for nurturing a developing embryo. When a fertilized ovum fails to show up, the body hits the biological reset button and sloughs off the uterine lining before building it up all over again for next month's reproductive roulette. But that's nothing compared to what the female body does when a fertilized egg actually settles in for a nine-month stay. Strap yourselves in for a tour of the incredible female baby-making machinery — practice questions included.

Identifying the Female Reproductive Parts and Their Functions

First and foremost in the female reproductive repertoire are the two *ovaries,* which usually take a turn every other month to produce a single ovum. Roughly the size and shape of large unshelled almonds, the female gonads lie on either side of the uterus, below and slightly behind the *Fallopian tubes* (also called the *uterine tubes*). Each ovary has a *stroma* (body) of connective tissue surrounded by a dense fibrous connective tissue called the *tunica albuginea* (literally "white covering"); yes, that's the same name as the tissue surrounding the testes. In fact, the ovaries in a female and the testes in a male are *homologous,* meaning that they share similar origins. External to the tunica albuginea is a layer of cuboidal cells known as the *germinal epithelium.* During growth of the ovary in a female fetus prior to birth, the germinal epithelium dips into the body of the ovary in various places. Over time, a mass of epithelial cells called *primordial follicles,* or *primary follicles,* becomes separated from the main body of the ovary. The ovaries of a young girl contain from 100,000 to 400,000 of these follicles, most of them present at birth.

At puberty and approximately once each month until menopause (which we cover in the later section "Growing, Changing, and Aging"), the following happens:

1. **The *pars distal* (anterior lobe) of the *hypophysis* (pituitary gland) secretes *follicle-stimulating hormones*, or *FSH*, which prompt about 1,000 of the primordial follicles to resume cellular division by meiosis.**

   Usually, only one follicle matures to become a *Graafian follicle* (twins, triplets, or even more fetuses result if more than one follicle matures to the point of releasing an ovum).

2. **One cell of this mass, the *oocyte* (produced by oogenesis, or meiosis), becomes the ovum while the remaining cells surround the ovum as part of the *cumulus oophorus* and others line the fluid-filled follicular cavity as the *membrana granulosa*.**

3. **As the ovum matures, its follicle moves toward the ovary's surface and begins secreting the hormone *estrogen*, which signals the *endometrium* (uterine lining) to build up in preparation for pregnancy.**

4. **As blood levels of estrogen begin to rise, the pituitary stops releasing FSH and begins releasing *luteinizing hormone*, or *LH*, which prompts the Graafian follicle now at the surface of the ovary to rupture, triggering *ovulation* — the release of the ovum.**

5. **Ringed by follicular cells in what's called the *corona radiata*, the ovum enters the *coelom* (body cavity) and is swept into the *Fallopian tube* by a fringe of tissue called *fimbriae*.**

6. **It takes approximately three days for the ovum to travel from the Fallopian tube to the uterus.**

Meanwhile, back at the ovary, a clot has formed inside the ruptured follicle and the membrana granulosa cells are being replaced by yellow luteal cells, forming a *corpus luteum* (literally "yellow body") on the surface of the ovary. This new endocrine gland secretes *progesterone,* a hormone that signals the uterine lining to prepare for possible implantation of a fertilized egg, inhibits the maturing of Graafian follicles, ovulation, and the production of estrogen to prevent menstruation; and stimulates further growth in the mammary glands (which is why some women get sore breasts a few days before their periods begin). If pregnancy occurs, the placenta also will release progesterone to prevent menstruation throughout the pregnancy.

If the ovum isn't fertilized, the corpus luteum dissolves after 10 to 14 days to be replaced by scar tissue called the *corpus albicans.* If pregnancy does occur, the corpus luteum remains and grows for about six months before disintegrating. Only about 400 of a woman's primordial follicles ever get a chance to make the trip to the uterus. The rest ripen to various stages before degenerating into what are known as *atretic follicles* (or *corpora atretica*) over the course of her lifetime. Figure 14-1 shows what happens inside an ovary.

Fallopian tubes, oviducts, uterine tubes — call them what you will, but they're where the real business of fertilization takes place. Why? Because an egg must be fertilized within 24 hours of its release from the ovary to remain viable. These small, muscular tubes lined with *cilia* are nearly 5 inches long and, somewhat surprisingly, aren't directly connected to the ovaries. Instead, the funnel-shaped end, the *infundibulum*, of a tube is fragmented into finger-like projections called *fimbriae* that help to move the ovum from the body cavity into the tube. When it's in the tube, where fertilization takes place, the combined motions of both cilia and peristalsis (the same muscle contractions that move food through the digestive system) propel the ovum toward the uterus for implantation. If a fertilized egg implants anywhere else — say in the hollow of the Fallopian tube itself — the pregnancy is referred to as *ectopic* and the woman must have immediate surgery to remove the developing embryo before it can damage any vital organs.

While not attached to the ovaries, the Fallopian tubes are attached to the pear-shaped *uterus,* which is located between the urinary bladder and the rectum. Its upper, wide end is called the *fundus;* the lower, narrow end that opens into the vagina is the *cervix;* and the central region is the *body.* Endometrium lines the uterus in varying amounts depending on the stage of a woman's menstrual cycle or pregnancy. This lining is supported by a thick muscular layer called the *myometrium,* which is under the control of the autonomic nervous system and comes into play when the uterus contracts, such as during labor.

Sperm enter and menstrual fluid leaves through the *vagina,* a muscular tube that connects the uterus with the outside of the body. Lined with a fold of highly elastic mucous membrane, the vagina can enlarge greatly during childbirth. A folded membrane of connective tissue called the *hymen* lies at the opening of the vaginal canal until it is ruptured or torn, often by sexual intercourse but sometimes by other physical activities. At either side of the vaginal opening are two *Bartholin glands* that secrete a lubricating mucous.

On the outside, the female genitalia extends toward the posterior from a mound of soft, fatty tissue called the *mons pubis* that covers the bone structure called the *pubic symphysis.* Behind this, the *vulva* consists of two flaps of fatty tissue: the outer lips, or *labia majora;* and the smaller, hairless inner lips, the *labia minora.* Just above where the inner lips join is a small flap of tissue called the *clitoral hood,* under which is the *clitoris,* erectile tissue that swells during sexual arousal. Below the clitoris is the external opening of the urethra and below that is the *introitus,* the opening to the vagina. You can see the female reproductive system illustrated in Figure 14-2.

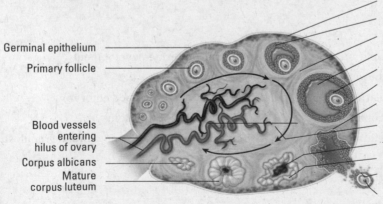

Germinal epithelium

Primary follicle

Blood vessels entering hilus of ovary

Corpus albicans

Mature corpus luteum

Tunica albuginea

Secondary follicle

Cortex of stroma

Follicular fluid

Cumulus oophorus

Graafian follicle

Membrana granulosa

Medulla of stroma

Blood clot

Early corpus luteum

Corona radiata

Ovulation results in discharged ovum

Figure 14-1: The ovary.

Illustration by Imagineering Media Services Inc.

Find out how familiar you are with the female anatomy:

1. Follicular growth (or change) in the ovary is started by

   a. Progesterone

   b. Estrogen

   c. LH

   d. FSH

   e. Testosterone

2. Which one of the following is not part of the maturing Graafian follicle?

   a. Follicular fluid

   b. Ovum

   c. Membrana granulosum

   d. Corpora atretica

3. Which one of the following is not a function of estrogen?

   a. Preparing the endometrium

   b. Supporting development of the ovum

   c. Supporting development of the corpus luteum

   d. Preventing secretion of FSH from the pituitary

4. Which of the following does not produce a hormone?

   a. Corpus luteum

   b. Pars distalis

   c. Graafian follicle

   d. Corpus albicans

   e. Placenta

5.–9. Match the term to its description.

5. _____ Ectopic	**a.** Period of intrauterine development
6. _____ Gestation	**b.** Cessation of menses
7. _____ Ovulation	**c.** Development of out-of-place embryo
8. _____ Menopause	**d.** Release of ovum into coelom
9. _____ Luteinization	**e.** Glandular development by membrana granulosa

10. Progesterone is produced by the

   a. Endometrium

   b. Pars distalis

   c. Follicle

   d. Corpus luteum

   e. Pituitary

11.–15. Match the term to its description.

11. _____ Corona radiata   **a.** Endocrine gland that secretes progesterone

12. _____ Endometrium   **b.** Lining of the follicle

13. _____ Corpus luteum   **c.** Follicle cells surrounding the ovum

14. _____ Stroma   **d.** Body of the ovary

15. _____ Membrana granulosa   **e.** Inner lining of the uterus

16.–19. Match the description to the hormone (one answer is used twice).

16. _____ Prevents menstruation in pregnant females   **a.** Progesterone

17. _____ Male sex hormone   **b.** Testosterone

18. _____ Secretion of the developing follicle   **c.** Estrogen

19. _____ Secreted by the corpus luteum

20.–35. Use the terms that follow to identify the anatomy of the female reproductive system shown in Figure 14-2.

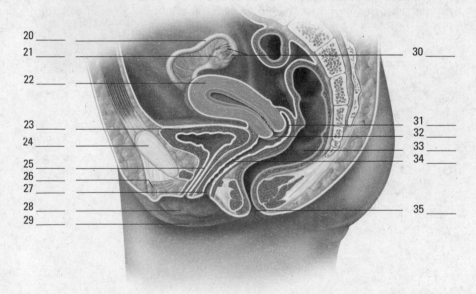

Figure 14-2:
The female reproductive system.

Illustration by Imagineering Media Services Inc.

a. Cervix   **i.** Vaginal orifice

b. Fimbriae   **j.** Anus

c. Urinary bladder   **k.** Symphysis pubis

d. Clitoris   **l.** Uterus

e. Labium major   **m.** Labium minor

f. Ovary   **n.** Uterine tube

g. Rectum   **o.** Vagina

h. Posterior fornix   **p.** Urethra

Making Eggs: A Mite More Meiosis

We cover meiosis in detail in Chapter 13, but we also cover the process in this section, exploring how meiosis contributes to making a *zygote* (fertilized egg) and its chromosomal makeup. Normal human diploid, or 2N, cells contain 23 pairs of homologous chromosomes for a total of 46 chromosomes each. One chromosome from each pair comes from the individual's mother, and the other comes from the father. Each homologous pair contains the same type of genetic information, but the *expression* of this genetic information may differ from one chromosome of the pair to the homologous chromosome. For example, one chromosome from the mother may carry the genetic coding for blue eyes, whereas the homologous chromosome from the father may code for brown eyes.

Although all the body's cells have 46 chromosomes — including the cells that eventually mature into the ovum and the sperm — each gamete (either ovum or sperm) must offer only half that number if fertilization is to succeed. That's where meiosis steps in. As you see in Figure 14-3, the number of chromosomes is cut in half during the first meiotic division, producing gametes that are haploid, or 1N. The second meiotic division produces four haploid sperms in the male but only one functional haploid ovum in the female. When fertilization occurs, the new zygote will contain 23 pairs of homologous chromosomes for a total of 46 chromosomes. The zygote then proceeds through mitosis to produce the body's cells, distributing copies of all 46 chromosomes to each new cell.

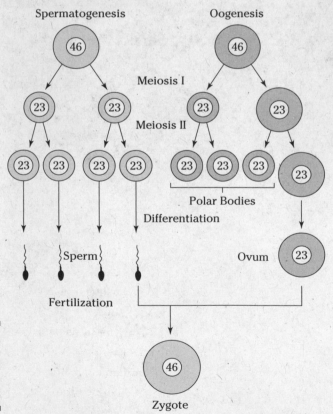

Figure 14-3: How chromosomes divide up in meiosis.

Think you've conquered this process? Find out by answering the following practice questions:

Q. How many chromosomes are in most of the cells in your body?

   a. 46

   b. 23

   c. 92

A. The correct answer is 46. That's the usual human complement.

36. How many chromosomes are in each sperm produced by a male?

   a. 46

   b. 23

   c. 92

37. How many chromosomes are in each ovum produced by a female?

   a. 46

   b. 23

   c. 92

38. How many chromosomes are in a newly fertilized egg, or zygote?

   a. 46

   b. 23

   c. 92

39. A haploid cell contains how many chromosomes?

   a. 46

   b. 23

   c. 92

40. The process of chromosomal reduction is called

   a. Mitosis

   b. Menses

   c. Meiosis

   d. Gene expression

41.–48. Fill in the blanks to complete the following sentences:

Meiosis produces sperm and ovum, which contributes to making a **41.** _____ (fertilized egg) and its chromosomal makeup. Normal humans have **42.** _____, or 2N, cells containing 23 pairs of homologous chromosomes for a total of 46 chromosomes each. A pair of chromosomes containing the same type of genetic information are **43.** _____ chromosomes. An ovum and a sperm are called sex cells or **44.** _____. The number of chromosomes is cut in half during the first meiotic division, producing gametes that are **45.** _____, or 1N. The second meiotic division produces **46.** _____ sperms in the male but only **47.** _____ovum in the female. The zygote then proceeds through **48.** _____ to produce the body's cells.

Making Babies: An Introduction to Embryology

Fertilization (the joining of ovum and sperm) occurs in the Fallopian tube as the ovum travels toward the uterus. Fertilization must occur within 24 hours of ovulation or the ovum degenerates. But that doesn't mean that sexual intercourse outside that time frame can't lead to pregnancy. Research indicates that spermatozoa can survive up to seven days inside the uterus and Fallopian tubes. If a sperm is still motile — that is, if it's still whipping its flagellum tail — when an ovum comes down the tube, it will do what it was made to do and penetrate the ovum's membrane.

When the sperm penetrates the ovum, it releases enzymes that allow it to digest its way into the ovum, leaving its flagellum behind. After that first sperm penetrates, the membrane instantly solidifies around the ovum, preventing any other sperm from getting inside. Presto! You've got a zygote. Over the next three to five days, the zygote moves through the Fallopian tube to the uterus, undergoing *cleavage* (mitotic cell division) along the way: Two cells become four smaller cells, four cells become eight smaller cells, and then those eight cells become a solid 16-cell ball called a *morula*. After five days of cleavage, the cells form a hollow ball of approximately 32 cells called a *blastula,* or *blastocyst*. The inner hollow region is called the *blastocoele,* and the outer-layer cells are called the *trophoblast.* Figure 14-4 illustrates this process of development.

Within three days of its arrival in the uterine cavity (generally within a week of fertilization), the blastocyst implants in the endometrium, and some of the blastocyst's cells — called *totipotent embryonic stem cells* — organize into an inner cell mass called the *embryonic disk,* or *embryoblast.* Over time, the embryonic disk *differentiates* into the tissues of the developing embryo (see Figure 14-4). Cells above the disk form the *amniotic cavity,* and those below form the *gut cavity* and two primitive *germ layers.* The layer nearest the amniotic cavity forms the *ectoderm* while that nearest the gut cavity forms the *endoderm.* Between the two layers, additional ectodermal cells develop to form a third layer, the *mesoderm.* The ectoderm forms skin and nerve tissue; the mesoderm forms bones, cartilage, connective tissue, muscles, and organs; and the endoderm forms the linings of the organs and glands.

To keep these terms straight, remember that *endo–* means "inside or within," *ecto–* means "outer or external," and *meso–* means "middle."

After three weeks of development, the heart begins beating. In the fourth week of development, the embryonic disk forms an elongated structure that attaches to the developing placenta by a connecting stalk. A head and jaws form while primitive buds sprout; the buds will develop into arms and legs. During the fifth through seventh weeks, the head grows rapidly and a face begins to form (eyes, nose, and a mouth). Fingers and toes grow at the ends of the elongating limb buds. All internal organs have started to form. After eight weeks of development, the embryo begins to have a more human appearance and is referred to as a *fetus.*

The outer cells of the embryo, together with the endometrium of the uterus, form the *placenta,* a new internal organ that exists only during pregnancy. The placenta attaches the fetus to the uterine wall, exchanges gases and waste between the maternal and fetal bloodstreams, and secretes hormones to sustain the pregnancy.

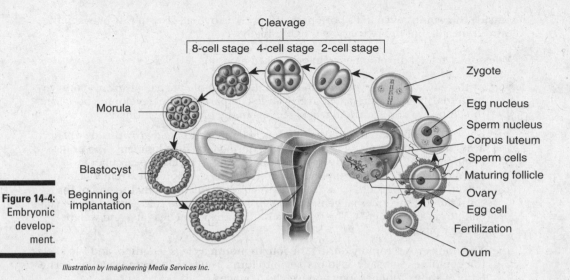

Figure 14-4: Embryonic development.

Cleavage

8-cell stage   4-cell stage   2-cell stage

Zygote

Egg nucleus

Sperm nucleus

Corpus luteum

Sperm cells

Maturing follicle

Ovary

Egg cell

Fertilization

Ovum

Morula

Blastocyst

Beginning of implantation

Illustration by Imagineering Media Services Inc.

Now that you've refreshed your memory a bit about how babies are made (beyond the birds and bees talks), try the following practice questions:

49.–58. Mark the statement with a T if it's true or an F if it's false:

49. _____ The mesoderm will form nerve tissue and skin.

50. _____ The embryonic stage is completed at the end of the eighth week.

51. _____ Cells above the embryonic disk form the gut cavity.

52. _____ During the fifth through seventh weeks, the arm and leg buds elongate and fingers and toes begin to form.

53. _____ Cleavage is successive mitotic divisions of the embryonic cells into smaller and smaller cells.

54. _____ As the zygote moves through the uterine (Fallopian) tube, it undergoes meiosis.

55. _____ The placenta serves to exchange gases and waste between the maternal blood and the fetal blood.

56. _____ After five days of cleavage, the cells form into a hollow ball called the morula.

57. _____ The embryonic stage is completed at the end of the fifth week of development.

58. _____ Sexual intercourse five days before ovulation cannot lead to pregnancy.

Growing from Fetus to Baby

Pregnancy is divided into three periods called *trimesters* (although many new parents bemoan a postnatal fourth trimester until the baby sleeps through the night). The first 12 weeks of development mark the first trimester, during which *organogenesis* (organ formation) is established. During the second trimester, all fetal systems continue to develop and rapid growth triples the fetus's length. By the third trimester, all organ systems are functional and the fetus usually is considered *viable* (capable of surviving outside the womb) even if it's born prematurely. The overall growth rate slows in the third trimester, but the fetus gains weight rapidly.

Milestones in fetal development can be marked monthly.

- ✔ At the **end of the second month** (when the terminology changes from "embryo" to "fetus"), the head remains overly large compared to the developing body, and the limbs are still short. All major regions of the brain have formed.

- ✔ During the **third month,** body growth accelerates and head growth slows. The arms reach the length they will maintain during fetal development. The bones begin to ossify, and all body systems have begun to form. The circulatory (cardiovascular) system supplies blood to all the developing extremities, and even the lungs begin to practice "breathing" amniotic fluid. By the end of the third month, the external genitalia are visible in the male (ultrasound technicians call this a "turtle sign"). The fetus is a bit less than 4 inches long and weighs about 1 ounce.

- ✔ The body grows rapidly during the **fourth month** as legs lengthen, and the skeleton continues to ossify as joints begin to form. The face looks more human. The fetus is about 7 inches long and weighs 4 ounces.

- ✔ Growth slows during the **fifth month,** and the legs reach their final fetal proportions. Skeletal muscles become so active that the mother can feel fetal movement. Hair grows on the head, and *lanugo,* a profusion of fine soft hair, covers the skin. The fetus is about 12 inches long and weighs ½ to 1 pound.

- ✔ The fetus gains weight during the **sixth month,** and eyebrows and eyelashes form. The skin is wrinkled, translucent, and reddish because of dermal blood vessels. The fetus is between 11 and 14 inches long and weighs a bit less than 1½ pounds.

- ✔ During the **seventh month,** skin becomes smoother as the fetus gains subcutaneous fat tissue. The eyelids, which are fused during the sixth month, open. Usually, the fetus turns to an upside-down position. It's between 13 and 17 inches long and weighs from 2½ to 3 pounds.

- ✔ During the **eighth month,** subcutaneous fat increases, and the fetus shows more baby-like proportions. The testes of a male fetus descend into the scrotum. The fetus is now 16 to 18 inches long and has grown to just under 5 pounds.

- ✔ During the **ninth month,** the fetus plumps up considerably with additional subcutaneous fat. Much of the lanugo is shed, and fingernails extend all the way to the tips of the fingers. The average newborn at the end of the ninth month is 20 inches long and weighs about 7½ pounds.

The following practice questions deal with the development of the fetus during its 40 weeks in the womb:

59. A fetus usually is considered viable

  a. By the second trimester

  b. By the third trimester

  c. By the fifth month of gestation

  d. By the 19th week of gestation

60. Describe one new fetal development for each month:

3rd month: _____

4th month: _____

5th month: _____

6th month: _____

7th month: _____

8th month: _____

9th month: _____

Growing, Changing, and Aging

After a baby arrives, the female reproductive system goes into a different form of overdrive. Throughout the pregnancy, the placenta has been producing estrogen and progesterone to sustain the fetus. But after the baby is born, the sudden drop in hormonal blood levels triggers the pituitary gland to release *prolactin,* a hormone that stimulates the woman's mammary glands to secrete milk in a process called *lactation.* First, however, the glands produce *colostrum,* a thin, yellowish fluid rich in antibodies and minerals to sustain a newborn. Both colostrum and milk flow from a number of lobes inside the breast through *lactiferous ducts* that converge on the nipple.

From birth to 4 weeks of age, the newborn is called a *neonate.* Faced with survival after its physical separation from the mother, a neonate must abruptly begin to process food, excrete waste, obtain oxygen, and make circulatory adjustments.

From 4 weeks to 2 years of age, the baby is called an *infant.* Growth during this period is explosive under the stimulation of circulatory growth hormones from the pituitary gland, adrenal steroids, and thyroid hormones. The infant's *deciduous* teeth, also called *baby* or *milk teeth,* begin to form and erupt through the gums. The nervous system advances, making coordinated activities possible. The baby begins to develop language skills.

From 2 years to puberty, you're looking at a *child.* Influenced by growth hormones, growth continues its rapid pace as deciduous teeth are replaced by permanent teeth. Muscle coordination, language skills, and intellectual skills also develop rapidly.

From puberty, which starts between the ages of 11 and 14, to adulthood, the child is called an *adolescent.* Growth occurs in spurts. Girls achieve their maximum growth rate between the ages of 10 and 13, whereas boys experience their fastest growth between the ages of 12 and 15. Primary and secondary sex characteristics begin to appear. Growth terminates when the epiphyseal plates of the long bones ossify sometime between the ages of 18 and 21. Motor skills and intellectual abilities continue to develop, and psychological changes occur as adulthood approaches.

The young adult stage covers roughly 20 years, from the age of 20 to about 40. Physical development reaches its peak and adult responsibilities are assumed, often including a career, marriage, and a family. After about age 30, physical changes that indicate the onset of aging begin to occur.

From age 40 to about 65, physiological aging continues. Gray hair, diminished physical abilities, and skin wrinkles are outward signs of aging. Women go through *menopause,* the cessation of monthly cycles, which is also known as *climacteric* or the *change of life.* While *menarche* (the onset of menstruation) may begin any time between 10 and 15 years of age, the female body's monthly reproductive cycle slows and stops entirely between the ages of 40 and 55. With the cessation of menses comes a decrease in size of the uterus, shortening of the vagina, shrinkage of the mammary glands, disappearance of Graafian follicles, and shrinkage of the ovaries. For about six years prior to menopause, many women experience a stage called *perimenopause* during which increasingly irregular hormone secretions can cause fluctuations in menstruation and a sensation called *hot flashes*.

From age 65 until death is the period of *senescence,* the process of aging. Individual adults can show widely varying patterns of aging in part because of differences in genetic background and physical activities. Signs of senescence include loss of skin elasticity and accompanying sagging or wrinkling; weakened bones and decreasingly mobile joints; weakened muscles; impaired coordination, memory, or intellectual function; cardiovascular problems; reduced immune responses; decreased respiratory function caused by reduced lung elasticity; and decreased peristalsis and muscle tone in the digestive and urinary tracts.

Following are some practice questions on the aging cycle:

61. The gland involved in lactation is called the

a. Pituitary gland

b. Mammary gland

c. Pars distalis

d. Adrenal gland

e. Parathyroid gland

62. An individual's height stops increasing

a. During adolescence

b. Upon onset of menopause

c. When the epiphyseal plates of the long bones ossify

d. During middle age

63. A 20-month-old child is formally called a(n)

a. Baby

b. Infant

c. Toddler

d. Neonate

64.–70. Match the description to its life stage.

64. _____ Assume adult responsibilities, possibly including marriage and a family

65. _____ Faced with survival, it must process food, excrete waste, and obtain oxygen

66. _____ Primary and secondary sex characteristics begin to appear

67. _____ Experiences the period of senescence

68. _____ Deciduous teeth begin to form

69. _____ Women go through menopause

70. _____ From 2 years of age to puberty

a. Neonate
b. Infant
c. Child
d. Adolescent
e. Young adult
f. Middle-aged adult
g. Old adult

Answers to Questions on the Female Reproductive System

The following are answers to the practice questions presented in this chapter.

1 Follicular growth (or change) in the ovary is started by **d. FSH.** That's short for follicle stimulating hormone, which is released by the pituitary gland.

2 Which one of the following is not part of the maturing Graafian follicle? **d. Corpora atretica.** Follicles that never matured all the way become corpora atretica over time.

3 Which one of the following is not a function of estrogen? **c. Supporting development of the corpus luteum.** By the time the corpus luteum starts to develop, estrogen already has begun to bow out of the reproductive equation.

4 Which of the following does not produce a hormone? **d. Corpus albicans.** Corpus albicans is the scar tissue that follows disintegration of the corpus luteum, so it makes sense that it doesn't produce any hormones.

5 Ectopic: **c. Development of out-of-place embryo.** *Ektopos* in Greek literally means "out of place."

6 Gestation: **a. Period of intrauterine development.** "Gestation" is just a fancy term for pregnancy, and *intra*– means "within."

7 Ovulation: **d. Release of ovum into coelom.** Ignore the reference to the body cavity. "Release of ovum" is the giveaway here.

8 Menopause: **b. Cessation of menses.** Menses is the same as menstruation.

9 Luteinization: **e. Glandular development by membrana granulosa.** The granulosa cells become luteal cells to form the corpus luteum, which is actually a new gland each month.

10 Progesterone is produced by the **d. corpus luteum.**

11 Corona radiata: **c. Follicle cells surrounding the ovum**

12 Endometrium: **e. Inner lining of the uterus**

13 Corpus luteum: **a. Endocrine gland that secretes progesterone**

14 Stroma: **d. Body of the ovary**

15 Membrana granulosa: **b. Lining of the follicle**

16 Prevents menstruation in pregnant females: **a. Progesterone**

17 Male sex hormone: **b. Testosterone**

18 Secretion of the developing follicle: **c. Estrogen**

19 Secreted by the corpus luteum: **a. Progesterone**

20–35 Following is how Figure 14-2, the female reproductive system, should be labeled.

> 20. **n. Uterine tube**; 21. **f. Ovary**; 22. **l. Uterus**; 23. **c. Urinary bladder**; 24. **k. Symphysis pubis**; 25. **p. Urethra**; 26. **d. Clitoris**; 27. **i. Vaginal orifice**; 28. **m. Labium minor**; 29. **e. Labium major**; 30. **b. Fimbrae**; 31. **h. Posterior fornix**; 32. **a. Cervix**; 33. **g. Rectum**; 34. **o. Vagina**; 35. **j. Anus**

36 How many chromosomes are in each sperm produced by a male? **b. 23.** Cut the usual complement in half to form a gamete.

37 How many chromosomes are in each ovum produced by a female? **b. 23.** Same answer for the same reason as in the preceding question.

38 How many chromosomes are in a newly fertilized egg, or zygote? **a. 46.** The 23 in the ovum plus the 23 in the sperm equals 46.

39 A haploid cell contains how many chromosomes? **b. 23.**

The number of chromosomes in a haploid cell may be easier to remember this way: *Haploid* is *half,* but *diploid* is *double* that.

40 The process of chromosomal reduction is called **c. meiosis.** Don't be fooled by this question; mitosis is when a cell replicates itself — with a full complement of chromosomes. Meiosis is also cell division, but it occurs only when the body is working to make gametes for reproduction.

41–48 Meiosis produces sperm and ovum, which contributes to making a **41. zygote** (fertilized egg) and its chromosomal makeup. Normal humans have **42. diploid,** or 2N, cells containing 23 pairs of homologous chromosomes for a total of 46 chromosomes each. A pair of chromosomes containing the same type of genetic information are **43. homologous** chromosomes. An ovum and a sperm are called sex cells or **44. gametes.** The number of chromosomes is cut in half during the first meiotic division, producing gametes that are **45. haploid,** or 1N. The second meiotic division produces **46. four haploid** sperms in the male but only **47. one functional haploid** ovum in the female. The zygote then proceeds through **48. mitosis** to produce the body's cells.

49 The mesoderm will form nerve tissue and skin. **False**

50 The embryonic stage is completed at the end of the eighth week. **True**

51 Cells above the embryonic disk form the gut cavity. **False**

52 During the fifth through seventh weeks, the arm and leg buds elongate and fingers and toes begin to form. **True**

53 Cleavage is successive mitotic divisions of the embryonic cells into smaller and smaller cells. **True**

54 As the zygote moves through the uterine (Fallopian) tube, it undergoes meiosis. **False**

55 The placenta serves to exchange gases and waste between the maternal blood and the fetal blood. **True**

56 After five days of cleavage, the cells form into a hollow ball called the morula. **False**

57 The embryonic stage is completed at the end of the fifth week of development. **False**

58 Sexual intercourse five days before ovulation cannot lead to pregnancy. **False.** It certainly can happen — and does.

59 A fetus usually is considered viable **b. by the third trimester.** Even then, a premature birth can have serious health consequences for the newborn.

60 Describe one new fetal development for each month:

3rd month: Bones begin to ossify, body growth accelerates while head growth slows, lungs begin to practice breathing amniotic fluid, external genitalia visible in male, 4 inches long and about 1 ounce

4th month: Body grows rapidly, legs lengthen, joints begin to form, face looks more human, roughly 7 inches long and 4 ounces

5th month: Mother can feel fetal movement, hair grows on head, lanugo covers skin, about 12 inches long and 8 to 16 ounces

6th month: Eyebrows and eyelashes form, weight gain accelerates, roughly 11 to 14 inches long and just under 1 ½ pounds

7th month: Subcutaneous fat begins to form, eyelids open, usually turns to upside-down position, between 13 and 17 inches long and 2 ½ to 3 pounds

8th month: Subcutaneous fat increases, fetus appears more "baby-like," testes of male descend into scrotum, 16 to 18 inches long and roughly 5 pounds

9th month: Substantial "plumping" with subcutaneous fat, lanugo shed, fingernails extend to fingertips, average newborn is 20 inches long and weighs 7 ½ pounds

61 The gland involved in lactation is called the **b. mammary gland.** Everybody say "moo"!

62 An individual's height stops increasing **c. when the epiphyseal plates of the long bones ossify.** And that occurs sometime between the ages of 18 and 21.

63 A 20-month-old child is formally called a(n) **b. infant.** The terms "baby" and "toddler" aren't part of the formal medical lexicon.

64 Assume adult responsibilities, possibly including marriage and a family: **e. Young adult**

65 Faced with survival, it must process food, excrete waste, and obtain oxygen: **a. Neonate**

66 Primary and secondary sex characteristics begin to appear: **d. Adolescent**

67 Experiences the period of senescence: **g. Old adult**

68 Deciduous teeth begin to form: **b. Infant**

69 Women go through menopause: **f. Middle-aged adult**

70 From 2 years of age to puberty: **c. Child**

Part V
Mission Control: All Systems Go

The 5th Wave   By Rich Tennant

"Dave's thyroid is very active."

In this part . . .

This part reviews how the body you know so well moves, thinks, feels, and thrives on a day-to-day basis. You get to know the intricacies of the nerves and brain, arguably the most complex of all the anatomical systems as well as the fine-tuning capacity of the endocrine system and its hormones.

Chapter 15

Feeling Jumpy: The Nervous System

Throughout this book, you look at the human body from head to toe, exploring how it collects and distributes the molecules it needs to grow and thrive, how it reproduces itself, and even how it gets rid of life's nastier byproducts. In this chapter, however, you look at the living computer that choreographs the whole show, the one system that contributes the most to making us who we are as humans.

The nervous system is the communications network that goes into nearly every part of the body, enervating your muscles, pricking your pain sensors, and letting you reach beyond yourself into the larger world. More than 80 major nerves make up this intricate network, and each nerve contains somewhere around 1 million *neurons* (individual nerve cells). It's through this complex network that you respond both to external and internal stimuli, demonstrating a characteristic called *irritability* (the capacity to respond to stimuli, not the tendency to yell at annoying people).

There are three functional types of cells in the nervous system: *receptor cells* that receive a stimulus (sensing); *conductor cells* that transmit impulses (integrating); and *effector cells*, or motor neurons, which bring about a response such as contracting a muscle. Put another way, there are three functions of the human nervous system as a whole: *orientation,* or the ability to generate nerve impulses in response to changes in the external and internal environments (this also can be referred to as *perception*); *coordination,* or the ability to receive, sort, and direct those signals to channels for response (this also can be referred to as *integration*); and *conceptual thought,* or the capacity to record, store, and relate information received and to form plans for future reactions to environmental change (which includes specific *action*).

In this chapter, you get a feel for how the nervous system is put together. You practice identifying the parts and functions of nerves and the brain itself as well as the structure and activities of the Big Three parts of the whole nervous system: the central, the peripheral, and the autonomic systems. In addition, we touch on the sensory organs that bring information into the human body.

Building from Basics: Neurons, Nerves, Impulses, Synapses

Before trying to study the system as a whole, it's best to break it down into building blocks first.

Neurons

The basic unit that makes up nerve tissue is the *neuron* (also called a *nerve cell*). Its properties include that marvelous *irritability* that we speak of in the chapter introduction as well as *conductivity,* otherwise known as the ability to transmit a *nerve impulse.*

The central part of a neuron is the *cell body,* or *soma,* that contains a large nucleus with one or more nucleoli, mitochondria, Golgi apparatus, numerous ribosomes, and *Nissl bodies* that are associated with conduction of a nerve impulse. (See Chapter 2 for an overview of a cell's primary parts.) Extending from the cell body are threads of cytoplasm, or cytoplasmic projections, containing specialized fibrils, or *neurofibrillae.* Two types of cytoplasmic projections play a role in neurons: *Dendrites* conduct impulses to the cell body while *axons* (nerve fibers) usually conduct impulses away from the cell body (see Figure 15-1). Each neuron has only one axon; however, each axon can have many branches called *axon collaterals*, enabling communication with many target cells. The point of attachment on the soma is called the *axon hillock.* In addition, each neuron may have one dendrite, several dendrites, or none at all.

There are three types of neurons, as follows:

- **Motor neurons,** or *efferent neurons,* transmit messages from the brain and spinal cord to effector organs, including muscles and glands, triggering them to respond. Motor neurons are classified structurally as *multipolar* because they're star-shaped cells with a single large axon and numerous dendrites.

- **Sensory neurons,** or *afferent neurons,* are triggered by physical stimuli, such as light, and pass the impulses on to the brain and spinal cord. Sensory fibers have special structures called receptors, or end organs, where the stimulus is propagated. Sensory neurons can be classified structurally as either *monopolar* or *bipolar.* Monopolar neurons have a single *process* (a projection or outgrowth of tissue) that divides shortly after leaving the cell body; one branch conveys impulses from sense organs while the other branch carries impulses to the central nervous system. Bipolar neurons have two processes — one dendrite and one axon.

- **Association neurons** (also called *internuncial neurons, interneurons,* or *intercalated neurons*) are triggered by sensory neurons and relay messages between neurons within the brain and spinal cord. Interneurons, like motor neurons, are classified structurally as multipolar.

Here are a couple of handy memory devices: Afferent connections arrive, and efferent connections exit. Dendrites deliver impulses while axons send them away.

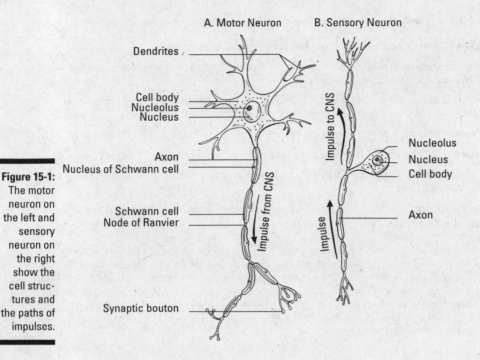

A. Motor Neuron   B. Sensory Neuron

Dendrites

Cell body
Nucleolus
Nucleus

Impulse to CNS

Nucleolus
Nucleus
Cell body

Axon
Nucleus of Schwann cell

Impulse from CNS

Schwann cell
Node of Ranvier

Impulse

Axon

Synaptic bouton

Figure 15-1:
The motor neuron on the left and sensory neuron on the right show the cell structures and the paths of impulses.

Nerves

Whereas neurons are the basic unit of the nervous system, *nerves* are the cable-like bundles of axons that weave together the peripheral nervous system. There are three types of nerves:

- **Afferent nerves** are composed of sensory nerve fibers (axons) grouped together to carry impulses from receptors to the central nervous system.

- **Efferent nerves** are composed of motor nerve fibers carrying impulses from the central nervous system to effector organs, such as muscles or glands.

- **Mixed nerves** are composed of both afferent and efferent nerve fibers.

The diameter of individual axons (nerve fibers) tends to be microscopically small — many are no more than a micron, or one-millionth of a meter. But these same axons extend to lengths of 1 millimeter and up. The longest axons in the human body run from the base of the spine to the big toe of each foot, meaning that these single-cell fibers may be 1 meter or more in length.

Each axon is swathed in *myelin,* a white fatty material made up of concentric layers of *Schwann cells* in peripheral nerves. *Oligodendrocytes* in the central nervous system are also associated with myelinated nerve fibers. The result is a structure referred to as a *myelin sheath.* Gaps in the sheath called *nodes of Ranvier* give the underlying nerve fiber access to extracellular fluid, to speed up propagation of the nerve impulse. *Nonmyelinated nerve fibers* lie within body organs and therefore don't need protective myelin sheaths to help them transmit impulses. Many peripheral nerve cell fibers also are protected by a *neurilemmal sheath,* a membrane that surrounds both the nerve fiber and its myelin sheath.

From the inside out, nerves are composed of the following:

- **Axon:** The impulse-conducting process of a neuron
- **Myelin sheath:** An insulating envelope that protects the nerve fiber and facilitates transmission of nerve impulses
- **Neurolemma (or neurilemma):** A thin membrane present in many peripheral nerves that surrounds the nerve fiber and the myelin sheath
- **Endoneurium:** Loose, or *areolar*, connective tissue surrounding individual fibers
- **Fasciculi:** Bundles of fibers within a nerve
- **Perineurium:** The same kind of connective tissue as endoneurium; surrounds a bundle of fibers
- **Epineurium:** The same kind of connective tissue as endoneurium and perineurium; surrounds several bundles of fibers

There also is a class of cells called *neuroglia,* or simply *glia,* that act as the supportive cells of the nervous system, providing neurons with nutrients and otherwise protecting them. Glia include oligodendrocytes that support the myelin sheath within the central nervous system; star-shaped cells called *astrocytes* that both support nerve tissue and contribute to repairs when needed; and *microgliacytes,* cells that remove dead or dying parts of tissue (this type of cell is called a *phagocyte,* which literally translates from the Greek words for "cell that eats").

Impulses

Neuron membranes are *semi-permeable* (meaning that certain small molecules like ions can move in and out but larger molecules can't), and they're *electrically polarized* (meaning that positively charged ions called *cations* rest around the outside membrane surface while negatively charged ions called *anions* line the inner surface; you can find more about ions in Chapter 1).

A neuron that isn't busy transmitting an impulse is said to be at its *resting potential.* But the *nerve impulse theory,* or *membrane theory,* says that things switch around when a stimulus — a nerve impulse, or *action potential* — moves along the neuron. A stimulus changes the specific permeability of the fiber membrane and causes a depolarization due to a reshuffling of the cations and anions. This change spreads along the nerve fiber and constitutes the nerve impulse. It's called an *all-or-none response* because each neuron has a specific threshold of excitation. Once that threshold is exceeded, the nerve fiber responds with a fixed impulse. After depolarization, repolarization occurs followed by a refractory period, during which no further impulses occur, even if the stimuli's intensity increases.

Intensity of sensation, however, depends on the frequency with which one nerve impulse follows another and the rate at which the impulse travels. That rate is determined by the diameter of the impacted fiber and tends to be more rapid in large nerve fibers. It's also more rapid in myelinated fibers than nonmyelinated fibers. The cytoplasm of the axon or nerve fiber is electrically conductive and the myelin decreases the capacitance to prevent charge leakage through the membrane. Depolarization at one node of Ranvier is sufficient to trigger regeneration of the voltage at the next node. Therefore, in myelinated nerve fibers the action potential does not move as a wave but recurs at successive nodes, traveling faster than in nonmyelinated fibers. This is referred to as saltatory conduction (from the Latin word *saltare,* which means "to hop or leap").

Synapses

Neurons don't touch, which means that when a nerve impulse reaches the end of a neuron, it needs to cross a gap to the next neuron or to the gland or muscle cell for which the message is intended. That gap is called a *synapse,* or *synaptic cleft.* An electric synapse — generally found in organs and glial cells — uses channels known as *gap junctions* to permit direct transmission of signals between neurons. But in other parts of the body, chemical changes occur to let the impulse make the leap. The end branches of an axon each form a terminal knob or bulb called a *bouton terminal* (that first word's pronounced *boo-taw*), beyond which there is a space between it and the next nerve pathway. When an impulse reaches the bouton terminal, the following happens:

1. *Synaptic vesicles* in the knob release a transmitter called *acetylcholine* that flows across the gap and increases the permeability of the next cell membrane in the chain.

2. An enzyme called *cholinesterase* breaks the transmitter down into *acetyl* and *choline,* which then diffuse back across the gap.

3. An enzyme called *choline acetylase* in the synaptic vesicles reunites the acetyl and choline, prepping the bouton terminal to do its job again when the next impulse rolls through.

Nervous about getting all this right? Try some practice questions:

1.–5. Match the term to its description.

1. _____ Irritability

2. _____ Conductivity

3. _____ Orientation

4. _____ Coordination

5. _____ Conceptual thought

a. Tissue's ability to respond to stimulation

b. Ability to receive impulses and direct them to channels for favorable response

c. Sense organs' capacity to generate nerve impulse to stimulation

d. Spreading of the nerve impulse

e. Capacity to record, store, and relate information to be used to determine future action

6. The brain and spinal cord are called the

a. Central nervous system

b. Visceral afferent system

c. Autonomic nervous system

d. Peripheral nervous system

7. The functional unit of the nervous system is the

a. Axon

b. Nephron

c. Dendron

d. Neuron

8. The terminal structure of the cytoplasmic projection of the neuron *cannot* be a(n)

   a. Node of Ranvier

   b. End organ

   c. Effector

   d. End bulb

   e. Receptor

9. The afferent fiber that carries impulses to a neuron's cell body is called a

   a. Nissl body

   b. Neuron

   c. Dendrite

   d. Axon

   e. Mitochondria

10. The membrane surrounding the axon (nerve fiber) is the

   a. Sarcolemma

   b. Neurilemma

   c. Perineurium

   d. Epineurium

11.–15. Match the term to its description.

   11. _____ Astrocytes

   12. _____ Microgliacytes

   13. _____ Oligodendrocytes

   14. _____ Axons

   15. _____ Dendrites

   a. Cytoplasmic projections carrying impulses to the cell body

   b. Cells that form and preserve myelin sheaths

   c. Cytoplasmic projections carrying impulses from the cell body

   d. Cells that are phagocytic

   e. Cells that contribute to the repair process of the central nervous system

16. The neuroglia cells are important as

   a. Sensory tissue

   b. Supporting tissue

   c. Irritable tissue

   d. Conducting tissue

17. Axons tend to consist of

   a. Single processes

   b. Several synapses

   c. Multiple processes

   d. None of the above

   e. Both a and c

18. A synapse between neurons is best described as the

    a. Transmission of a continuous impulse

    b. Transmission of an electrical impulse

    c. Transmission of an impulse through a chemical and physical change

    d. Transmission of an impulse through a physical change

    e. Transmission of an impulse through a chemical change

19.–23. Match the term to its description.

    19. _____ Endoneurium

    20. _____ Neurilemma

    21. _____ Schwann cell

    22. _____ Node of Ranvier

    23. _____ Myelin sheath

    a. Fatty layer around an axon fiber

    b. Outer thin membrane around an axon fiber

    c. Cell in the sheath of an axon

    d. Depression in the sheath around a fiber

    e. Connective tissue surrounding individual fibers in a nerve

24.–28. Match the term to its description.

    24. _____ All-or-none response

    25. _____ Cation

    26. _____ Anion

    27. _____ Polarization

    28. _____ Depolarization

    a. Impermeability of cell membrane

    b. Negatively charged ion on the inner surface of the cell membrane

    c. Threshold of excitation determines ability to respond

    d. Positively charged ion on the outer surface of the cell membrane

    e. Reshuffling of cell membrane ions; permeability of cell membrane

29.–33. Match the term to its description.

    29. _____ Cholinesterase

    30. _____ Choline acetylase

    31. _____ Terminal bulb

    32. _____ Acetylcholine

    33. _____ Synapse

    a. Excitatory chemical necessary for continual nerve pathway

    b. Enzyme for breakdown of excitatory chemical

    c. Enzyme for reformation of excitatory chemical

    d. Space between neurons

    e. Contains storage vesicles for excitatory chemical

Minding the Central Nervous System and the Brain

Together, the brain and spinal cord make up the central nervous system. The spinal cord, which forms very early in the embryonic spinal canal, extends down into the tail portion of the vertebral column. But because bone grows much faster than nerve tissue, the end of the cord soon is too short to extend into the lowest reaches of the spinal canal. In an adult, the 18-inch spinal cord ends between the first and second lumbar vertebrae, roughly where the last ribs attach. Its tapered end is called the *conus medullaris*. The cord continues as separate strands below that point and is referred to as the *cauda equina* (horse tail). A thread of fibrous tissue called the *filum terminale* extends to the base of the *coccyx* (tailbone) and is attached by the coccygeal ligament.

Spinal cord

An oval-shaped cylinder with two deep grooves running its length at the back and the front, the spinal cord doesn't fill the spinal cavity by itself. Also packed inside are the *meninges,* cerebrospinal fluid, a cushion of fat, and various blood vessels.

Three membranes called *meninges* envelop the central nervous system, separating it from the bony cavities. The *dura mater,* the outer layer, is the hardest, toughest, and most fibrous layer and is composed of white collagenous and yellow elastic fibers. The *arachnoid,* or middle membrane, forms a web-like layer just inside the dura mater. The *pia mater,* a thin inner membrane, lies close along the surface of the central nervous system. The pia mater and arachnoid may adhere to each other and are considered as one, called *pia-arachnoid.*

There are spaces or cavities between the pia mater and the arachnoid where major regions join, for instance where the medulla oblongata and the cerebellum join. These sub-arachnoid spaces are called *cisterna.* Spaces or cavities between the arachnoid layer and the dura mater layer are referred to as *subdural.*

Two types of solid material make up the inside of the cord, which you can see in Figure 15-2: *gray matter* (which is indeed grayish in color) containing unmyelinated neurons, dendrites, cell bodies, and neuroglia; and *white matter,* so-called because of the whitish tint of its myelinated nerve fibers. At the cord's midsection is a small *central canal* surrounded first by gray matter in the shape of the letter H and then by white matter, which fills in the areas around the H pattern. The legs of the H are called anterior, posterior, and lateral *horns* of gray matter, or *gray columns.*

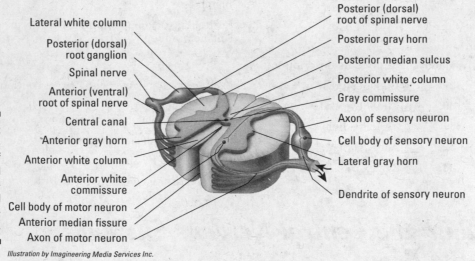

Figure 15-2:
A cross-section of the spinal cord, showing spinal nerve connections.

Lateral white column
Posterior (dorsal) root ganglion
Spinal nerve
Anterior (ventral) root of spinal nerve
Central canal
Anterior gray horn
Anterior white column
Anterior white commissure
Cell body of motor neuron
Anterior median fissure
Axon of motor neuron

Posterior (dorsal) root of spinal nerve
Posterior gray horn
Posterior median sulcus
Posterior white column
Gray commissure
Axon of sensory neuron
Cell body of sensory neuron
Lateral gray horn
Dendrite of sensory neuron

Illustration by Imagineering Media Services Inc.

The white matter consists of thousands of myelinated nerve fibers arranged in three *funiculi* (columns) on each side of the spinal cord that convey information up and down the cord's tracts. Ascending afferent (sensory) nerve tracts carry impulses to the brain; descending efferent (motor) nerve tracts carry impulses from the brain. Each tract is named according to its origin and the joint of synapse, such as the corticospinal and spinothalmic tracts.

Thirty-one pairs of spinal nerves arise from the sides of the spinal cord and leave the cord through the *intervertebral foramina* (spaces) to form the peripheral nervous system, which we discuss in the later section "Taking Side Streets: The Peripheral Nervous System."

Brain

One of the largest organs in the adult human body, the brain tips the scales at 3 pounds and packs roughly 100 billion neurons (yes, that's billion with a "b") and 900 billion supporting neuroglia cells. In this section, we review six major divisions of the brain from the bottom up (see Figure 15-3): *medulla oblongata, pons, midbrain, cerebellum, diencephalon,* and *cerebrum.*

Medulla oblongata

The spinal cord meets the brain at the *medulla oblongata,* or *brainstem,* just below the right and left cerebellar hemispheres of the brain. In fact, the medulla oblongata is continuous with the spinal cord at its base (inferiorly) and back (dorsally) and located anteriorly and superiorly to the pons. All the afferent and efferent tracts of the cord can be found in the brainstem as part of two bulges of white matter forming an area referred to as the *pyramids.* Many of the tracts cross from one side to the other at the pyramids, which explains why the right side of the brain controls the left side of the body and vice versa.

Along with the pons, the medulla oblongata also forms a network of gray and white matter called the *reticular formation,* the upper part of the so-called *extrapyramidal pathway.* With its capacity to arouse the brain to wakefulness, it keeps the brain alert, directs messages in the form of impulses, monitors stimuli entering the sense receptors (accepting some and rejecting others it deems to be irrelevant), refines body movements, and effects higher mental processes such as attention, introspection, and reasoning. Although the cortex of the cerebrum is the actual powerhouse of thought, it must be stimulated into action by signals from the reticular formation.

Nerve cells in the brainstem are grouped together to form nerve centers (nuclei) that control bodily functions, including cardiac activities, and respiration as well as reflex activities such as sneezing, coughing, vomiting, and alimentary tract movements. The medulla oblongata affects these reactions through the vagus, also referred to as cranial nerve X or the 10th cranial nerve. Three other cranial nerves also originate from this area: the 9th (IX) or glossopharyngeal, 11th (XI) or accessory, and 12th (XII) or hypoglossal.

Pons

The *pons* (literally "bridge") does exactly as its name implies: It connects the cerebellum through a structure called the middle peduncle, the cerebrum by the superior peduncle, and the medulla oblongata by the inferior peduncle. It also unites the cerebellar hemispheres, coordinates muscles on both sides of the body, controls facial muscles (including those used to chew), and regulates the first stage of respiration. Oh, and it contains the nuclei for the following cranial nerves: the 5th (V) or trigeminal, the 6th (VI) or abducens, 7th (VII) or facial, and 8th (VIII) or vestibulocochlear.

Midbrain

Between the pons and the diencephalon lies the *mesencephalon,* or *midbrain.* It contains the *corpora quadrigemina,* which correlates optical and tactile impulses as well as regulates muscle tone, body posture, and equilibrium through reflex centers in the *superior colliculus.* The *inferior colliculus* contains auditory reflex centers and is believed to be responsible for the detection of musical pitch. The midbrain contains the *cerebral aqueduct,* which connects the third ventricle of the *thalamus* with the fourth ventricle of the medulla oblongata (see the section "Ventricles" later in this chapter for more). The mesencephalon contains nuclei for the 3rd (III) or oculomotor cranial nerve and the 4th (IV) or trochlear cranial nerve. The red nucleus that contains *fibers* of the *rubrospinal tract,* a motor tract that acts as a relay station for impulses from the cerebellum and higher brain centers, also lies within the midbrain, constituting the superior cerebellar peduncle.

Cerebellum

The *cerebellum* also is known as the *little brain* or *small brain.* The second-largest division of the brain, it's just above and overhangs the medulla oblongata and lies just beneath the rear portion of the cerebrum. Inside, the cerebellum resembles a tree called the *arbor vitae,* or "tree of life." A central body called the *vermis* connects the two lateral masses called the *cerebellar hemispheres* and assists in motor coordination and refinement of muscular movement, aiding equilibrium and muscle tone. The cerebellar cortex or gray matter contains Purkinje neurons with pear-shaped cell bodies, a multitude of dendrites, and a single axon. It sends impulses to the white matter of the cerebellum and to other deeper nuclei in the cerebellum, and then to the brainstem. The cerebellar cortex has parallel ridges called the folia cerebelli, which are separated by deep sulci.

Diencephalon

The *diencephalon,* a region between the mesencephalon and the cerebrum, contains separate brain structures called the *thalamus, epithalamus, subthalamus,* and *hypothalamus.* The region where the two sides of the thalami come in contact and join forces is called the *intermediate mass.* The thalamus is a primitive receptive center through which the sensory impulses travel on their way to the cerebral cortex. Here, nerve fibers from the spinal cord and lower parts of the brain synapse with neurons leading to the sensory areas of the cortex of the cerebrum. The thalamus is the great integrating center of the brain with the ability to correlate the impulses from tactile, pain, olfactory, and gustatory (taste) senses with motor reactions.

The epithalamus contains the *choroid plexus,* a vascular structure that produces spinal fluid. The pineal body and olfactory centers also lie within the epithalamus, which forms the roof of the third ventricle. The *subthalamus* is located below the thalamus and regulates the muscles of emotional expression.

The *hypothalamus* contains the centers for sexual reflexes; body temperature; water, carbohydrate, and fat metabolism; and emotions that affect the heartbeat and blood pressure. It also has the *optic chiasm* (connecting the optic nerves to the optic tract), the posterior lobe of the pituitary gland, and a funnel-shaped region called the *infundibulum* that forms the stalk of the pituitary gland.

Cerebrum

The *cerebrum,* or forebrain, is often called the *true brain.* It has two cerebral hemispheres — the right and the left. A thin outer layer of gray matter called the *cerebral cortex* features folds or convolutions called *gyri;* furrows and grooves are referred to as *sulci,* and deeper grooves are called *fissures.* A longitudinal fissure separates the cerebrum. The transverse fissure separates the cerebrum and the cerebellum. Each hemisphere has a set of controls for sensory and motor activities of the body. Interestingly, it's not just right-side/left-side controls that are reversed in the cerebrum; the upper areas of the cerebral cortex control the lower body activities while the lower areas of the cortex control upper-body activities in a reversal called "little man upside down."

Commissural fibers, a tract of nerves running from one side of the brain to the other, coordinate activities between the right and left hemispheres. The *corpus callosum* physically unites the two hemispheres and is the largest and densest mass of commissural fibers. A smaller mass called the *fornix* also plays a role.

Different functional areas of the cerebral cortex are divided into lobes:

- **Frontal lobe:** The seat of intelligence, memory, and idea association
- **Parietal lobe:** Functions in the sensations of temperature, touch, and sense of position and movement as well as the perception of size, shape, and weight
- **Temporal lobe:** Is responsible for perception and correlation of acoustical stimuli
- **Occipital lobe:** Handles visual perception

Medulla

The medulla, the region interior to the cortex, is composed of white matter that consists of three groups of fibers. *Projection fibers* carry impulses afferently from the brain stem to the cortex and efferently from the cortex to the lower parts of the central nervous system. *Association fibers* originate in the cortical cells and carry impulses to the other areas of the cortex on the same hemisphere. *Commissural fibers* connect the two cerebral hemispheres.

Ventricles

The brain's four *ventricles* are cavities and canals filled with cerebrospinal fluid. Two lateral ventricles are separated by the *septum pellucidum*. The lateral ventricles communicate with the third ventricle through the *foramen of Monro*. The third ventricle is connected by the *cerebral aqueduct* to the fourth ventricle, which is continuous with the central canal of the spinal cord and contains openings to the meninges. The fourth ventricle has openings that allow fluid to enter into the subarachnoid spaces.

Lining the ventricles is a thin layer of epithelial cells known as *ependyma,* or the *ependymal layer.* Along with a network of capillaries from the pia mater, the ependyma and capillaries form the *choroid plexus,* which is the source of cerebrospinal fluid. The choroid plexus of each lateral ventricle produces the greatest amount of fluid. Fluid formed by the choroid plexus filters out by *osmosis* (refer to Chapter 2) and circulates through the ventricles. Fluid is returned to the blood through the *arachnoid villi,* finger-like projections of the *arachnoid meninx,* which absorbs the fluid.

Twelve pairs of cranial nerves connect to the central nervous system via the brain (as opposed to the 31 pairs that connect via the spinal cord). Cranial nerves are identified by Roman numerals I through XII, and memorizing them is a classic test of anatomical knowledge. Check out Table 15-1 for a listing of all the nerves, and then read on for a memory tool.

Table 15-1		Cranial Nerves	
Number	**Name**	**Type**	**Function**
I	Olfactory	Sensory	Smell
II	Optic	Sensory	Vision
III	Oculomotor	Mixed nerve	Eyeball muscles
IV	Trochlear	Mixed nerve	Eyeball muscles

(continued)

Table 15-1 *(continued)*

Number	Name	Type	Function
V	Trigeminal	Tri means "three," so the three types of trigeminal nerves are 1) Opthalmic nerve: sensory nerve; skin and mucous membranes of face and head; 2) Maxillary nerve: mixed nerve; mastication; 3) Mandibular nerve: mixed nerve; mastication	Skin; mastication (chewing)
VI	Abducens	Mixed nerve	Eye movements
VII	Facial	Mixed nerve	Facial expression; salivary secretion; taste
VIII	Vestibulocochlear	Sensory	Auditory nerve for hearing and equilibrium
IX	Glossopharyngeal	Mixed nerve	Taste; swallowing muscles of pharynx
X	Vagus	Mixed nerve	Controls most internal organs (viscera) from head and neck to transverse colon
XI	Accessory	Mixed nerve	Swallowing and phonation
XII	Hypoglossal	Motor nerve	Tongue movements

The first letters of each of these nerve names, in order, are OOOTTAFVGVAH. That's a mouthful, but students have come up with a number of memory tools to remember them. Our favorite is: Old Opera Organs Trill Terrific Arias For Various Grand Victories About History.

Put your knowledge of the central nervous system to the test:

EXAMPLE

Q. The meninges' functions are primarily

  a. Immunological

  b. Supportive

  c. Protective

  d. Both a and b

  e. Both b and c

A. The correct answer is supportive and protective. Yes, meninges have two functions.

34. The cerebrum consists of two major halves called

   a. Cerebellar hemispheres

   b. Cerebral spheres

   c. Cerebellar spheres

   d. Cerebral hemispheres

35. The cerebrum is divided into two major halves by the

   a. Lateral fissure

   b. Transverse fissure

   c. Longitudinal fissure

   d. Fissure of Sylvius

   e. Central sulcus

36. In the cerebrum, the

   a. Right side tends to control the left side of the body and vice versa

   b. Upper area controls lower-body activity

   c. Lower area controls upper-body activity

   d. None of the above are correct

   e. A, b, and c are correct

37. The functions of the occipital lobe of the cerebrum pertain principally to

   a. Visual activity

   b. Autonomic control

   c. Associative reasoning

   d. Motor coordination

   e. Auditory control

38. The cerebellum functions primarily as a center of

   a. Visual activity

   b. Associative reasoning

   c. Auditory activity

   d. Autonomic coordination

   e. Motor control

39.–43. Match the term to its description.

   39. _____ White matter

   40. _____ Reticular formation

   41. _____ Funiculus

   42. _____ Dorsal root ganglion

   43. _____ Gray matter

   a. Has the capacity to arouse the brain to wakefulness

   b. Myelinated fibers

   c. Bundles of nerve fibers arranged in tracts

   d. Collection of cell bodies outside of the central nervous system

   e. Unmyelinated fibers, cell bodies, and neuroglia

44.-48. Match the term to its description.

44. _____ Pons

45. _____ Cerebellum

46. _____ Medulla oblongata

47. _____ Cerebrum

48. _____ Mesencephalon

a. Bridge connecting the medulla oblongata and cerebellum

b. Contains the centers that control cardiac, respiratory, and vasomotor functions

c. Contains the corpora quadrigemina and nuclei for the oculomotor and trochlear nerves

d. Controls motor coordination and refinement of muscular movement

e. Controls sensory and motor activity of the body

49. The largest quantity of cerebrospinal fluid originates from the

a. Foramen of Monro

b. Arachnoid villi

c. Lateral ventricle

d. Optic chiasm

e. Foramen of Luschka

50. The part of the brain that contains the thalamus, pituitary gland, and the optic chiasm is the

a. Diencephalon

b. Mesencephalon

c. Myelencephalon

d. Telencephalon

e. Metencephalon

51.–62. Use the terms that follow to identify the parts of the brain shown in Figure 15-3.

a. Pons

b. Thalamus

c. Cerebellum

d. Corpus callosum

e. Third ventricle

f. Hypothalamus

g. Cerebrum

h. Cerebral aqueduct

i. Midbrain

j. Pituitary gland

k. Medulla oblongata

l. Fourth ventricle

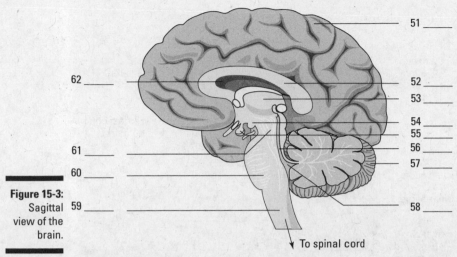

51 ____

62 ____   52 ____
                                           53 ____

                                           54 ____
                                           55 ____
61 ____   56 ____
60 ____   57 ____

Figure 15-3:
Sagittal 59 ____   58 ____
view of the
brain.

↓ To spinal cord

LifeART Image Copyright © 2007. Wolters Kluwer Health — Lippincott Williams & Wilkins

Taking Side Streets: The Peripheral Nervous System

The peripheral nervous system is the network that carries information to and from the spinal cord. Among its key structures are 31 pairs of spinal nerves (see Figure 15-4), each originating in a segment of the spinal cord called a *neuromere*. Eight of the spinal nerve pairs are *cervical* (having to do with the neck), 12 are *thoracic* (relating to the chest, or thorax), five are *lumbar* (between the lowest ribs and the pelvis), five are *sacral* (the posterior section of the pelvis), and one is *coccygeal* (relating to the tailbone). Spinal nerves connect with the spinal cord by two bundles of nerve fibers, or *roots*. The *dorsal root* contains afferent fibers that carry sensory information from receptors to the central nervous system. The cell bodies of these sensory neurons lie outside the spinal cord in a bulging area called the *dorsal root ganglion* (refer to the cross-section of the spinal cord in Figure 15-2). A second bundle, the *ventral root,* contains efferent motor fibers with cell bodies that lie inside the spinal cord. In each spinal nerve, the two roots join outside the spinal cord to form what's called a *mixed spinal nerve.*

Spinal reflexes, or *reflex arcs,* occur when a sensory neuron transmits a "danger" signal — like a sensation of burning heat — through the dorsal root ganglion. An internuncial neuron (or association neuron) in the spinal cord passes along the signal to a motor neuron (or efferent fiber) that stimulates a muscle, which immediately pulls the burning body part away from heat (see Figure 15-5).

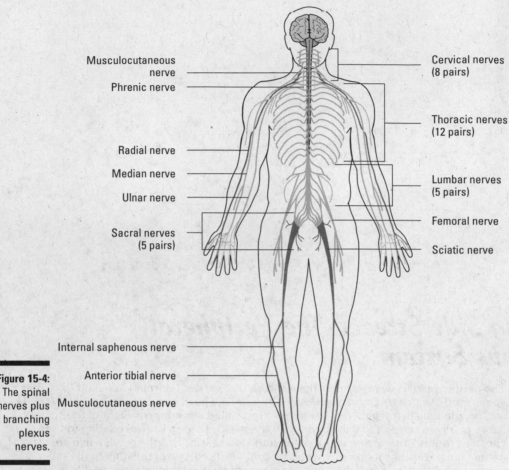

Figure 15-4: The spinal nerves plus branching plexus nerves.

Musculocutaneous nerve
Phrenic nerve
Radial nerve
Median nerve
Ulnar nerve
Sacral nerves (5 pairs)
Internal saphenous nerve
Anterior tibial nerve
Musculocutaneous nerve

Cervical nerves (8 pairs)
Thoracic nerves (12 pairs)
Lumbar nerves (5 pairs)
Femoral nerve
Sciatic nerve

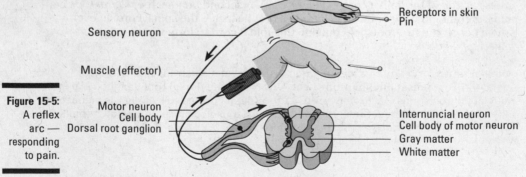

Figure 15-5: A reflex arc — responding to pain.

Sensory neuron
Muscle (effector)
Motor neuron
Cell body
Dorsal root ganglion

Receptors in skin
Pin
Internuncial neuron
Cell body of motor neuron
Gray matter
White matter

After a spinal nerve leaves the spinal column, it divides into two small branches. The posterior, or *dorsal ramus,* goes along the back of the body to supply a specific segment of the skin, bones, joints, and longitudinal muscles of the back. The ventral, or *anterior ramus,* is larger than the dorsal ramus and supplies the anterior and lateral regions of the trunk and limbs.

Groups of spinal nerves interconnect to form an extensive network called a *plexus* (Latin for "braid"), each of which connects through the anterior ramus, including the cervical plexus of the neck, brachial plexus of the arms, and lumbosacral plexus of the lower back (including the body's largest nerve, the *sciatic nerve*). However, there's no plexus in the thoracic area. Instead, the anterior ramus directly supplies the inter-costal muscles (literally "between the ribs") and the skin of the region.

63. The network of nerves formed by the ventral branch of the spinal nerve is a

   a. Pedicle

   b. Papilla

   c. Plexus

   d. Plica

   e. Phallus

64. How many spinal nerves are there?

   a. 61

   b. 31 pairs

   c. 13 pairs

   d. 1

   e. 12

65. Which spinal nerve area does not form a plexus?

   a. Cervical

   b. Lumbar

   c. Sacral

   d. Thoracic

Keep Breathing: The Autonomic Nervous System

Just as the name implies, the autonomic nervous system functions automatically. Divided into the *sympathetic* and *parasympathetic* systems, it activates the involuntary smooth and cardiac muscles and glands to serve such vital systems that function auto-matically as the digestive tract, circulatory system, respiratory, urinary, and endocrine systems. Autonomic functions are under the control of the hypothalamus, cerebral cortex, and medulla oblongata. The sympathetic system, which is responsible for the body's involuntary fight-or-flight response to stress, is defined by the autonomic fibers that exit the thoracic and lumbar segments of the spinal cord. The parasympathetic system is defined by the autonomic fibers that either exit the brainstem via the cranial nerves or exit the sacral segments of the spinal cord.

The sympathetic and parasympathetic systems oppose each other in function, helping to maintain *homeostasis,* or balanced activity in the body systems. Yet, often the sympa-thetic and parasympathetic systems work in concert. The sympathetic system dilates the eye's pupil, but the parasympathetic system contracts it again. The sympathetic system quickens and strengthens the heart while the parasympathetic slows the heart's action. The sympathetic system contracts blood vessels in the skin so more blood goes to muscles for a fight-or-flight reaction to stress, and the parasympathetic system dilates the blood vessels when the stress concludes.

As shown in Figure 15-6, a pair of sympathetic trunks lies to the right and left of the spinal cord and is composed of a series of ganglia that form nodular cords extending from the base of the skull to the front of the coccyx (tailbone). Sympathetic nerves originate as a short preganglionic neuron with its cell body inside the lateral horn of the gray matter of the spinal cord from the first thoracic to the third lumbar. Axons of these nerves then pass through the ventral root of the spinal nerve, leaving it through a branch of the spinal nerve called the *white rami* (named for their white myelin sheaths), which connect to one of the two chains of ganglia in the trunks. Within these hubs, synapses distribute the nerves to various parts of the body.

The parasympathetic system is referred to as a craniosacral system because its ganglia originate in the medulla oblongata (brainstem), mesencephalon, and the sacral portion of the spinal nerves, sending out impulses through the following cranial nerves: the oculomotor III, the facial VII, glossopharyngeal IX, and the vagus X. Parasympathetic nerves consist of long preganglionic fibers that synapse in a terminal ganglion near or within the organ or tissue that's being innervated. Generally speaking, the parasympathetic system acts in opposition to the sympathetic system.

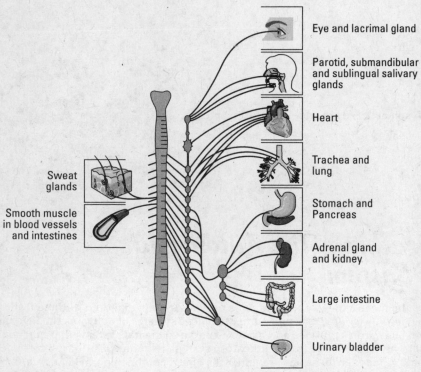

Sweat glands

Smooth muscle in blood vessels and intestines

Eye and lacrimal gland

Parotid, submandibular and sublingual salivary glands

Heart

Trachea and lung

Stomach and Pancreas

Adrenal gland and kidney

Large intestine

Urinary bladder

Figure 15-6: The sympathetic nervous system.

LifeART Image Copyright © 2007. Wolters Kluwer Health — Lippincott Williams & Wilkins

See whether any of the following practice questions touch a nerve:

66. The autonomic nervous system

   a. Innervates involuntary body functions

   b. Responds only at times of emotional stress

   c. Is composed only of sensory neurons

   d. Is separate anatomically and functionally from the cranial nerves

67. Which of the following statements is true about the autonomic nervous system?

   a. It has two parts: the parasympathetic that controls all normal functions, and the sympathetic that carries out the same functions.

   b. It's the nervous system that controls all reflexes.

   c. It doesn't function when the body's under stress.

   d. It has two divisions that are antagonistic to each other, meaning that one counteracts the effects of the other one.

   e. It controls the contractions of the skeletal, smooth, and cardiac muscle tissue.

68. The divisions of the autonomic nervous system are

   a. Sympathetic and peripheral

   b. Somatic and peripheral

   c. Parasympathetic and peripheral

   d. Sympathetic and parasympathetic

69. Which part of the autonomic nervous system can be called a craniosacral system?

   a. Ganglia

   b. Sympathetic trunks

   c. Parasympathetic system

   d. Medulla oblongata

Coming To Your Senses

The nervous system must have some way to perceive its environment in order to generate appropriate responses. That's where the senses come in. *Sense receptors* are those numerous organs that respond to stimuli — like increased temperature, bitter tastes, and sharp points — by generating a nerve impulse. While there are millions of *general sense receptors* found throughout the body that can convey touch, pain, and physical contact, there are far fewer of the *special sense receptors* — those located in the head — that really bring meaning to your world.

Sense receptors are classified by the stimuli they receive, as follows:

   ✔ **Exteroceptors:** Receive stimuli from the external environment. These are sensory nerve terminals, such as those in the skin and mucous membranes, that are stimulated by the immediate external environment.

   ✔ **Interoceptors:** Receive stimuli from the internal environment. These can be any of the sensory nerve terminals located in and transmitting impulses from the viscera.

   ✔ **Proprioceptors:** Part of the "true" internal environment. They're sensory nerve terminals chiefly found in muscles, joints, and tendons that give information concerning movements and position of the body.

   ✔ **Teleceptors:** Sensory nerve terminals stimulated by emanations from distant objects. They exist in the eyes, ears, and nose.

Eyes

Although there are many romantic notions about eyes, the truth is that an eyeball is simply a hollow sphere bounded by a trilayer wall and filled with a gelatinous fluid called, oddly enough, *vitreous humor* (see Figure 15-7). The outer fibrous coat is made up of the *sclera* in back and the *cornea* in front. The sclera provides mechanical support, protection, and a base for attachment of eye muscles, and it assists in the focusing process. The cornea covers the anterior with a clear window.

An intermediate, or vascular, coat called the *uvea* provides blood and lymphatic fluids to the eye, regulates light, and also secretes and reabsorbs *aqueous humor,* a thin watery liquid that fills the anterior chamber of the eyeball in front of the iris. A pigmented coat has three layers: the *iris,* containing blood vessels, pigment cells, and smooth muscle fibers to control the pupil's diameter; the *ciliary body,* which is attached to the periphery of the iris; and the *choroid,* a thin, dark brown, vascular layer lining most of the sclera on the back and sides of the eye. The choroid contains arteries, veins, and capillaries that supply the *retina* with nutrients, and it also contains pigment cells to absorb light and prevent reflection and blurring. An *optic nerve* enters at the back (posterior) of each eye.

The *retina* is part of an internal nervous layer that connects with the optic nerve. The nervous tissue layers along the inner back of the eye contain *rods* and *cones* (types of neurons that analyze visual input). The rods are dim light receptors whereas the cones detect bright light and construct form, structure, and color. The retina has an *optic disc,* which is essentially a blind spot incapable of producing an image.

The crystalline lens consists of concentric layers of protein. It's *biconcave* in shape, bulging outward. Located behind the pupil and iris, the lens is held in place by ligaments attached to the ciliary muscles. When the ciliary muscles contract, the shape of the lens changes, altering the visual focus. This process of *accommodation* allows the eye to see objects both at a distance and close up.

The *palpebrae* (eyelids) extend from the edges of the eye orbit, into which roughly five-sixths of the eyeball is recessed. Eyelids come together at medial and lateral angles of the eye that are called the *canthi.* In the medial angle of the eye is a pink region called the *caruncula,* or *caruncle.* The caruncula contains sebaceous glands and sudoriferous (sweat) glands. A mucous membrane called the *conjunctiva* covers the inner surface of each eyelid and the anterior surface of the eye. Up top and to the side of the orbital cavities are lacrimal glands that secrete tears that are carried through a series of lacrimal ducts to the conjunctiva of the upper eyelid. Ultimately, secretions drain from the eyes through the nasolacrimal ducts.

Ears

Human ears — otherwise called *vestibulocochlear* organs — are more than just organs of hearing. They also serve as organs of equilibrium, or balance. Here are the three divisions of the ear:

✔ The *external ear* includes the *auricle,* or *pinna,* which is the folded, rounded appendage made of cartilage and skin. Extending into the skull is the *ear canal,* or *external auditory meatus,* a short passage through the temporal bone ending at the *tympanic membrane,* or *eardrum.* Sebaceous glands near the external opening and *ceruminous* glands in the upper wall produce the brownish substance known as earwax, or *cerumen.*

✔ The *middle ear* is a small, usually air-filled cavity in the skull that's lined with mucous membrane. It communicates through the *Eustachian tube* with the pharynx. The Eustachian tube keeps air pressure equal on both sides of the eardrum (tympanic membrane), equalizing pressure in the middle ear with atmospheric pressure from outside. Three small bones called *auditory ossicles* occupy the middle ear, deriving their names from their shapes: the *malleus* (hammer), the *incus* (anvil), and the *stapes* (stirrup).

✔ The *internal ear* is the most complex structure of the entire organ because it's where vibrations are translated. It's composed of a group of interconnected canals or channels called the *cochlea*. Within the cochlea are three canals separated from each other by thin membranes; two of the canals — the *vestibular* and the *tympanic* — are bony chambers filled with a *perilymph fluid*, and the third canal — the *cochlear* canal — is a membranous chamber filled with *endolymph*. The cochlear canal lies between the vestibular and tympanic canals and contains the *organ of Corti*, a spiral-shaped organ made up of cells with projecting hairs that transmit auditory impulses.

The process of hearing a sound follows these basic steps:

1. **Sound waves travel through the auditory canal, striking the eardrum and making it vibrate and setting the three ossicle bones into motion.**

2. **The stapes at the end of the chain strikes against the oval window of the vestibular canal, translating the motion into the perilymph fluid in the vestibular and tympanic canals of the cochlea.**

3. **The vibrating fluid begins moving the basilar membrane that separates the two canals, stimulating the endolymph fluid in the membranous area of the cochlea.**

4. **The stimulated endolymph fluid in turn stimulates the hair cells of the organ of Corti, which transmit the impulses to the brain over the auditory nerve.**

That's the hearing part of your ears. Equilibrium requires that some additional parts come into play. Three semicircular canals, each with an *ampulla* (or small, dilated portion) at each end, lie at right angles to each other. The *ampullae* connect to a fluid-filled sac called a *utricle*, which in turn connects to another fluid-filled sac called a *saccule*. Both sacs contain regions called *maculae* that are lined with sensitive hairs and contain concretions (solid masses) of calcium carbonate called *otoliths* (or *otoconia*). When linear acceleration pulls at them, the otoliths press on the hair cells and initiate an impulse to the brain through basal sensory nerve fibers. When the head changes position, it causes a change in the direction of force on the hairs. Movement of the hairs stimulates dendrites of the *vestibulocochlear nerve* (the eighth cranial nerve) to carry impulses to the brain.

Q. The most sensitive region of the retina producing the greatest visual acuity is the

   a. Blind spot

   b. Cornea

   c. Fovea centralis

   d. Macula lutea

   e. Lens

A. The correct answer is fovea centralis. It's loaded with light-sensitive cones.

70.–83. Use the terms that follow to identify the internal structures of the eye shown in Figure 15-7.

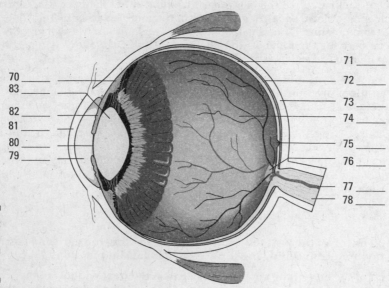

70
83

82
81
80
79

71
72
73
74
75
76
77
78

Figure 15-7:
The internal
structures
of the eye.

LifeART Image Copyright © 2007. Wolters Kluwer Health — Lippincott Williams & Wilkins

 a. Blind spot

 b. Pupil

 c. Optic nerve

 d. Retina

 e. Cornea

 f. Sclera

 g. Ciliary body

 h. Fovea centralis

 i. Lens

 j. Anterior cavity (aqueous humor)

 k. Choroid

 l. Blood vessels

 m. Iris

 n. Posterior cavity (vitreous humor)

84. The area of the eyeball that contains cells that are sensitive to light is the

 a. Cornea

 b. Retina

 c. Sclera

 d. Lens

 e. Optic nerve

85. Which of the following structures is not part of the eyeball?

    a. Optic nerve

    b. Iris

    c. Cornea

    d. Pupil

    e. Ciliary body

86. The accommodation (focusing) of the eye is accomplished by the

    a. Sphincter of the pupil

    b. Contraction of the iris

    c. Action of the ciliary muscles

    d. Dilator of the pupil

    e. Contraction of the pupil

87. The structure in the eye that responds to the ciliary muscles during focusing is the

    a. Pupil

    b. Lens

    c. Retina

    d. Iris

    e. Choroid

88. The middle ear is separated from the external ear by the

    a. Tympanic membrane

    b. Round window

    c. The organ of Corti

    d. Oval window

    e. Cochlea

89. The structure that contains the receptor cells for the perception of sound is the

    a. Tympanic membrane

    b. Semicircular canals

    c. Mastoid air cells

    d. Organ of Corti

    e. Middle ear cavity

90. The fluid in the membranous canal of the cochlea is called

    a. Aqueous humor

    b. Plasma

    c. Endolymph

    d. Perilymph

    e. Vitreous humor

91. Equilibrium is maintained by receptors in the

    a. Cochlea

    b. Utricle and saccule

    c. Tympanic membrane

    d. Organ of Corti

    e. Middle ear cavity

92. The small bone in the ear that strikes against the oval window of the vestibular canal, setting into motion the perilymph fluid in the vestibular and tympanic canals of the cochlea, is the

    a. Incus

    b. Hammer

    c. Malleus

    d. Anvil

    e. Stapes

93.–105. Use the terms that follow to identify the structures of the ear shown in Figure 15-8.

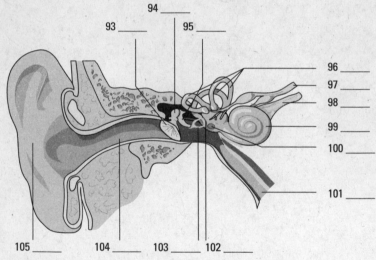

Figure 15-8: The anatomy of the ear.

LifeART Image Copyright © 2007. Wolters Kluwer Health — Lippincott Williams & Wilkins

    a. Oval window

    b. Semicircular canals

    c. Cochlea

    d. Pinna

    e. Malleus

    f. Cochlear nerve

    g. Stapes

    h. Incus

    i. Round window

    j. Auditory canal

    k. Tympanic membrane

    l. Vestibular nerve

    m. Auditory tube

Answers to Questions on the Nervous System

The following are answer to the practice questions presented in this chapter.

1 Irritability: **a. Tissue's ability to respond to stimulation**

2 Conductivity: **d. Spreading of the nerve impulse**

3 Orientation: **c. Sense organs' capacity to generate nerve impulse to stimulation**

4 Coordination: **b. Ability to receive impulses and direct them to channels for favorable response**

5 Conceptual thought: **e. Capacity to record, store, and relate information to be used to determine future action**

6 The brain and spinal cord are called the **a. central nervous system.** They're at the *center* of everything.

7 The functional unit of the nervous system is the **d. neuron.** Some of the other answer options are *parts* of a neuron, but the neuron is the central unit.

8 The terminal structure of the cytoplasmic projection of the neuron *cannot* be a(n) **a. node of Ranvier.** The nodes of Ranvier are gaps along the myelin sheath, so one of them can't be found at the end of the line.

9 The afferent fiber that carries impulses to a neuron's cell body is called a **c. dendrite.** Dendrites carry impulses to the neurons; axons carry them away.

10 The membrane surrounding the axon (nerve fiber) is the **b. neurilemma.** When present, this fiber actually wraps around the myelin sheath, so it's always on the outside.

11 Astrocytes: **e. Cells that contribute to the repair process of the central nervous system**

12 Microgliacytes: **d. Cells that are phagocytic**

13 Oligodendrocytes: **b. Cells that form and preserve myelin sheaths**

14 Axons: **c. Cytoplasmic projections carrying impulses from the cell body**

15 Dendrites: **a. Cytoplasmic projections carrying impulses to the cell body**

16 The neuroglia cells are important as **b. supporting tissue.** They nourish and protect neurons.

17 Axons tend to consist of **a. single processes.** A neuron may have one, many, or no dendrites, but it always has a single axon.

18 A synapse between neurons is best described as the **e. transmission of an impulse through a chemical change.** With all that acetylcholine and cholinesterase floating around, it must be a chemical transmission.

19 Endoneurium: **e. Connective tissue surrounding individual fibers in a nerve**

20 Neurilemma: **b. Outer thin membrane around an axon fiber**

21 Schwann cell: **c. Cell in the sheath of an axon**

22 Node of Ranvier: **d. Depression in the sheath around a fiber**

23 Myelin sheath: **a. Fatty layer around an axon fiber**

24 All-or-none response: **c. Threshold of excitation determines ability to respond**

25 Cation: **d. Positively charged ion on the outer surface of the cell membrane**

26 Anion: **b. Negatively charged ion on the inner surface of the cell membrane**

27 Polarization: **a. Impermeability of cell membrane**

28 Depolarization: **e. Reshuffling of cell membrane ions; permeability of cell membrane**

29 Cholinesterase: **b. Enzyme for breakdown of excitatory chemical**

30 Choline acetylase: **c. Enzyme for reformation of excitatory chemical**

31 Terminal bulb: **e. Contains storage vesicles for excitatory chemical**

32 Acetylcholine: **a. Excitatory chemical necessary for continual nerve pathway**

33 Synapse: **d. Space between neurons**

34 The cerebrum consists of two major halves called **d. cerebral hemispheres.** Cerebrum = cerebral, and two halves = hemispheres.

35 The cerebrum is divided into two major halves by the **c. longitudinal fissure.** Longitudinal is the most likely position for an equal division.

36 In the cerebrum, the **e. a, b, and c are correct (right side tends to control the left side of the body and vice versa, upper area controls lower-body activity, and lower area controls upper-body activity).** Right = left, and up = down. Clear as mud?

37 The functions of the occipital lobe of the cerebrum pertain principally to **a. visual activity.** To remember, use the word "occipital" to bring to mind the word "optic," which of course is related to visual activity.

38 The cerebellum functions primarily as a center of **e. motor control.**

39 White matter: **b. Myelinated fibers**

40 Reticular formation: **a. Has the capacity to arouse the brain to wakefulness**

41 Funiculus: **c. Bundles of nerve fibers arranged in tracts**

42 Dorsal root ganglion: **d. Collection of cell bodies outside of the central nervous system**

43 Gray matter: **e. Unmyelinated fibers, cell bodies, and neuroglia**

44 Pons: **a. Bridge connecting the medulla oblongata and cerebellum**

45 Cerebellum: **d. Controls motor coordination and refinement of muscular movement**

46 Medulla oblongata: **b. Contains the centers that control cardiac, respiratory, and vasomotor functions**

47 Cerebrum: **e. Controls sensory and motor activity of the body**

48 Mesencephalon: **c. Contains the corpora quadrigemina and nuclei for the oculomotor and trochlear nerves**

49 The largest quantity of cerebrospinal fluid originates from the **c. lateral ventricle.** This one requires rote memorization — sorry!

50 The part of the brain that contains the thalamus, pituitary gland, and the optic chiasm is the **a. diencephalon.** Think of it as the home of the thalamus, and you can't go wrong.

51–62 Following is how Figure 15-3, the brain, should be labeled.

> 51. **g. Cerebrum**; 52. **d. Corpus callosum**; 53. **b. Thalamus**; 54. **f. Hypothalamus**; 55. **j. Pituitary gland**; 56. **h. Cerebral aqueduct**; 57. **c. Cerebellum**; 58. **l. Fourth ventricle**; 59. **k. Medulla oblongata**; 60. **a. Pons**; 61. **i. Midbrain**; 62. **e. Third ventricle**

63 The network of nerves formed by the ventral branch of the spinal nerve is a **c. plexus.** The word stems from the Latin for "braid," which makes sense for a network.

64 How many spinal nerves are there? **b. 31 pairs.** Count them: 8 cervical, 12 thoracic, 5 lumbar, 5 sacral — plus 1 tailbone (coccygeal).

65 Which spinal nerve area does not form a plexus? **d. Thoracic**

66 The autonomic nervous system **a. innervates involuntary body functions.** It's the only answer option with a sense of automation.

67 Which of the following statements is true about the autonomic nervous system? **d. It has two divisions that are antagonistic to each other, meaning that one counteracts the effects of the other one.** As a result, the body achieves homeostasis.

68 The divisions of the autonomic nervous system are **d. sympathetic and parasympathetic.** They work against each other in order to help the body maintain balance.

69 Which part of the autonomic nervous system can be called a craniosacral system? **c. Parasympathetic system.** It originates in both the brainstem and the sacral region.

70–83 Following is how Figure 15-7, the internal structures of the eye, should be labeled.

> 70. **g. Ciliary body**; 71. **d. Retina**; 72. **k. Choroid**; 73. **f. Sclera**; 74. **n. Posterior cavity (vitreous humor)**; 75. **h. Fovea centralis**; 76. **a. Blind spot**; 77. **l. Blood vessels**; 78. **c. Optic nerve**; 79. **j. Anterior cavity (aqueous humor)**; 80. **b. Pupil**; 81. **e. Cornea**; 82. **m. Iris**; 83. **i. Lens**

84 The area of the eyeball that contains cells that are sensitive to light is the **b. retina.** It's at the back of the eyeball.

85 Which of the following structures is not part of the eyeball? **a. Optic nerve.** This nerve carries the visual signals to the brain.

86 The accommodation (focusing) of the eye is accomplished by the **c. action of the ciliary muscles.** They reshape the lens by contracting and relaxing as needed to bring things into focus.

87 The structure in the eye that responds to the ciliary muscles during focusing is the **b. lens.** Refer to the explanation for the preceding question.

88 The middle ear is separated from the external ear by the **a. tympanic membrane.** Otherwise known as the eardrum, this membrane sometimes bursts or tears as a result of infection or trauma.

89 The structure that contains the receptor cells for the perception of sound is the **d. organ of Corti.** Hairs in this structure are what ultimately send the signal down the auditory nerve.

90 The fluid in the membranous canal of the cochlea is called **c. endolymph.** Don't forget that the prefix *endo*– means "within."

91 Equilibrium is maintained by receptors in the **b. utricle and saccule.** These little endolymph-filled sacs have hairs and chunks of calcium carbonate that detect changes in gravitational forces.

92 The small bone in the ear that strikes against the oval window of the vestibular canal, setting into motion the perilymph fluid in the vestibular and tympanic canals of the cochlea, is the **e. stapes.** That's the only bone that actually touches the window. The other two carry the signal down the chain to the stapes.

93 – **105** Following is how Figure 15-8, the structures of the ear, should be labeled.

93. **k. Tympanic membrane**; 94. **e. Malleus**; 95. **h. Incus**; 96. **b. Semicircular canals**; 97. **l. Vestibular nerve**; 98. **f. Cochlear nerve**; 99. **c. Cochlea**; 100. **i. Round window**; 101. **m. Auditory tube**; 102. **a. Oval window**; 103. **g. Stapes**; 104. **j. Auditory canal**; 105. **d. Pinna**

Chapter 16

Raging Hormones:
The Endocrine System

The human body has two separate command and control systems that work in harmony most of the time but also work in very different ways. Designed for instant response, the nervous system cracks its cellular whip using electrical signals that make entire systems hop to their tasks with no delay (refer to Chapter 15). By contrast, the endocrine system's glands use chemical signals called *hormones* that behave like the steering mechanism on a large, fully loaded ocean tanker; small changes can have big impacts, but it takes quite a bit of time for any evidence of the change to make itself known. At times, parts of the nervous system stimulate or inhibit the secretion of hormones, and some hormones are capable of stimulating or inhibiting the flow of nerve impulses.

The word "hormone" originates from the Greek word *hormao,* which literally translates as "I excite." And that's exactly what hormones do. Each chemical signal stimulates some specific part of the body, known as *target tissues* or *target cells.* The body needs a constant supply of hormonal signals to grow, maintain homeostasis, reproduce, and conduct myriad processes.

In this chapter, we go over which glands do what and where, as well as review the types of chemical signals that play various roles in the body. You also get to practice discerning what the endocrine system does, how it does it, and why the body responds like it does.

No Bland Glands

Technically, there are ten or so primary endocrine glands with various other hormone-secreting tissues scattered throughout the body. Unlike *exocrine glands* (such as mammary glands and sweat glands), endocrine glands have no ducts to convey their secretions. Instead, hormones move directly into extracellular spaces surrounding the gland and from there move into capillaries and the greater bloodstream. Although they spread throughout the body in the bloodstream, hormones are uniquely tagged by their chemical composition. Thus they have separate identities and stimulate specific receptors on target cells so that usually only the intended cells or tissues respond to their signals.

All of the many hormones can be classified either as *steroid* (derived from cholesterol) or *nonsteroid* (derived from amino acids and other proteins). The steroid hormones — which include testosterone, estrogen, progesterone, and cortisol — are the ones most closely associated with emotional outbursts and mood swings. Steroidal hormones, which are nonpolar (see Chapter 2 for details on cell diffusion), penetrate cell membranes easily and initiate protein production at the nucleus.

Nonsteroid hormones are divided among four classifications:

✔ Some are derived from *modified amino acids,* including such things as epinephrine and norepinephrine, as well as melatonin.

✔ Others are *peptide*-based, including an antidiuretic hormone called ADH, oxytocin, and a melanocytes-stimulating hormone called MSH.

✔ *Glycoprotein*-based hormones include follicle-stimulating hormone (FSH), luteinizing hormone (LH), and chorionic gonadotropin — all closely associated with the female reproductive system.

✔ *Protein*-based nonsteroid hormones include such crucial substances as insulin and growth hormone as well as prolactin and parathyroid hormone.

Hormone functions include controlling the body's internal environment by regulating its chemical composition and volume, activating responses to changes in environmental conditions to help the body cope, influencing growth and development, enabling several key steps in reproduction, regulating components of the immune system, and regulating organic metabolism.

See if all this hormone-speak is sinking in:

1.–5. Mark the statement with a T if it's true or an F if it's false:

1. _____ The endocrine system brings about changes in the metabolic activities of the body tissue.

2. _____ The amount of hormone released is determined by the body's need for that hormone at the time.

3. _____ The glands of the endocrine system are composed of cartilage cells.

4. _____ Endocrine glands aren't functional in reproductive processes.

5. _____ Some hormones can be derivatives of amino acids, whereas others are synthesized from cholesterol.

6. Glands that secrete their product into the interstitial fluid, which flows into the blood, are

a. Exocrine glands

b. Endocrine glands

c. Heterocrine glands

d. Pericrintal glands

e. Interocrine glands

7. Cells that respond to a hormone are

a. Affectors

b. Effectors

c. Target cells

d. Chromosomal cells

e. Rickets cells

Mastering the Ringmasters

The key glands of the endocrine system include the *pituitary* (also called the *hypophysis*), *adrenal* (also referred to as *suprarenal*), *thyroid, parathyroid, thymus, pineal, islets of Langerhans* (within the *pancreas*), and *gonads* (testes in the male and ovaries in the female). But of all these, it's the pituitary working in concert with the hypothalamus in the brain that really keeps things rolling (see Figure 16-1).

The hypothalamus is the unsung hero linking the body's two primary control systems — the endocrine system and the nervous system. Part of the brain and part of the endocrine system, the hypothalamus is connected to the pituitary via a narrow stalk called the *infundibulum* that carries regular system status reports to the pituitary. In its supervisory role, the hypothalamus provides neurohormones to control the pituitary gland and influences food and fluid intake as well as weight control, body heat, and the sleep cycle.

The hypothalamus sits just above the pituitary gland, which is nestled in the middle of the human head in a depression of the skull's sphenoid bone called the *sella turcica.*

The pituitary's *anterior lobe,* also called the *adenohypophysis* or *pars distalis,* is sometimes called the "master gland" because of its role in regulating and maintaining the other endocrine glands. Hormones that act on other endocrine glands are called *tropic hormones;* all the hormones produced in the anterior lobe are polypeptides. Two capillary beds connected by venules make up the *hypophyseal portal system,* which connect the anterior lobe with the hypothalamus.

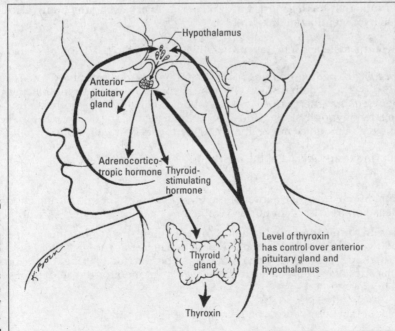

Figure 16-1:
The working relationship of the hypothalamus and the pituitary gland.

Hypothalamus

Anterior pituitary gland

Adrenocortico-tropic hormone

Thyroid-stimulating hormone

Thyroid gland

Level of thyroxin has control over anterior pituitary gland and hypothalamus

Thyroxin

Among the hormones produced in the anterior lobe of the pituitary gland are the following:

- **Follicle-stimulating hormone (FSH):** Signals an immature Graafian follicle in an ovary to mature into an ovum, which then produces the hormone estrogen. Negative feedback from the estrogen blocks further secretion of FSH. Guys, don't think you needn't worry about FSH: It's present in you, too, encouraging development and maturation of sperm.

- **Luteinizing hormone (LH):** Stimulates formation of the yellow body, or corpus luteum, on the surface of the ovary after an ovum has been released. In men, LH stimulates the development of interstitial cells and fresh production of testosterone.

- **Lactogenic hormone, or prolactin (PRL):** Promotes milk production in mammary glands, which are considered nonendocrine targets.

- **Interstitial-cell stimulating hormone (ICSH):** Stimulates formation and secretion of testosterone.

- **Thyrotropic hormone, or thyroid-stimulating hormone (TSH):** Controls the development and release of thyroid gland hormones *thyroxin* and *triiodothyronine*. The hypothalamus regulates TSH secretion by secreting thyrotropin-releasing hormone (TRH).

- **Adrenocorticotropic hormone (ACTH), or corticotropin:** Is a polypeptide composed of 39 amino acids that regulates the development, maintenance, and secretion of the cortex of the adrenal gland.

- **Somatotropic hormone, or growth hormone (GSH):** Stimulates body weight growth and regulates skeletal growth. This is the only hormone secreted by the anterior lobe that has a general effect on nearly every cell in the body (also regarded as nonendocrine targets).

For a review of the male and female reproductive systems, flip to Chapters 13 and 14.

The *posterior lobe*, or *neurohypophysis*, of the pituitary gland stores and releases secretions produced by the hypothalamus. This lobe is connected to the hypothalamus by the *hypophyseal tract*, nerve axons with cell bodies lying in the hypothalamus. Whereas the adenohypophysis is made up of epithelial cells, the neurohypophysis is largely composed of modified nerve fibers and neuroglial cells called *pituicytes*.

Among the hormones produced in the posterior lobe of the pituitary gland are the following:

- **Oxytocin:** Stimulates contraction of the uterine smooth muscle during childbirth and release of breast milk in nursing women

- **Vasopressin, or antidiuretic hormone (ADH):** Constricts smooth muscle tissue in the blood vessels, elevating blood pressure and increasing the amount of water reabsorbed by the kidneys, which reduces the production of urine. The hypothalamus has special neurons called *osmoreceptors* that monitor the amount of solute in the blood.

See how much of this information you're absorbing:

0. The hormone that stimulates ovulation is the

a. Follicle-stimulating hormone (FSH)

b. Antidiuretic hormone (ADH)

c. Oxytocin

d. Thyroid-stimulating hormone (TSH)

e. Luteinizing hormone (LH)

A. The correct answer is luteinizing hormone (LH). Don't be fooled into thinking it's FSH; that hormone does its job earlier, when it encourages an ovum to mature.

8.–12. Mark the statement with a T if it's true or an F if it's false:

8. _____ The pituitary gland consists of two parts: an endocrine gland and modified nerve tissue.

9. _____ The pituitary gland is found in the sella turcica of the temporal bone.

10. _____ The adenohypophysis is called the master gland because of its influence on all the body's tissues.

11. _____ ADH causes constriction of smooth muscle tissue in the blood vessels, which elevates the blood pressure.

12. _____ The neurohypophysis stores and releases secretions produced by the hypothalamus.

13. The gland that does the most to regulate and maintain the function of other glands is the

a. Pineal

b. Pituitary

c. Thyroid

d. Thymus

e. Parathyroid

14. Which of the following is not a pituitary hormone?

a. Progesterone

b. Follicle-stimulating hormone (FSH)

c. Growth hormone (GSH)

d. Prolactin

e. Luteinizing hormone (LH)

Supporting Cast of Glandular Characters

While the pituitary orchestrates the show at center stage, the endocrine system enjoys the support of a number of other important glands. Lying in various locations throughout the body, these glands secrete check-and-balance hormones that keep the body in tune.

Topping off the kidneys: The adrenal glands

Also called *suprarenals,* the adrenal glands lie atop each kidney. The central area of each is called the *adrenal medulla,* and the outer layers are called the *adrenal cortex.* Each glandular area secretes different hormones. The cells of the cortex produce over 30 steroids, including the hormones *aldosterone, cortisone,* and some sex hormones. The medullar cells secrete *epinephrine* (you may know it as *adrenaline*) and *norepinephrine* (also known as *noradrenaline*).

Made up of closely packed epithelial cells, the adrenal cortex is loaded with blood vessels. Layers form an outer, middle, and inner zone of the cortex. Each zone is composed of a different cellular arrangement and secretes different steroid hormones.

- The *zona glomerulosa* (outer zone) produces aldosterone.

- The *zona fasciculata* (middle zone) secretes cortisone (also called cortisol).

- The *zona reticularis* (inner zone) secretes small amounts of *gonadocorticoids* or sex hormones.

The following are among the hormones produced by the cortex:

- *Aldosterone,* or *mineralocorticoid,* regulates electrolytes (sodium and potassium mineral salts) retained in the body. It promotes the conservation of water and reduces urine output.

- *Cortisone,* or *cortisol,* acts as an antagonist to insulin, causing more glucose to form and increasing blood sugar to maintain normal levels. Elevated levels of cortisone speed up protein breakdown and inhibit amino acid absorption.

- *Androgens* and *estrogen* are cortical sex hormones. Androgens generally convey antifeminine effects, thus accelerating maleness, although in women adrenal androgens maintain the sexual drive. Too much androgen in females can cause *virilism* (male secondary sexual characteristics). Estrogen has the opposite effect, accelerating femaleness. Too much estrogen in a male produces feminine characteristics.

The adrenal medulla is made of irregularly shaped *chromaffin cells* arranged in groups around blood vessels. The sympathetic division of the autonomic nervous system controls these cells as they secrete adrenaline and noradrenaline. Both hormones have similar molecular structure and physiological functions. The adrenal cortex produces approximately 80 percent adrenaline and 20 percent noradrenaline. Adrenaline accelerates the heartbeat, stimulates respiration, slows digestion, increases muscle efficiency, and helps muscles resist fatigue. Noreadrenaline does similar things but also raises blood pressure by stimulating contraction of muscular arteries.

The terms "adrenaline" and "noradrenaline" are interchangeable with the terms "epinephrine" and "norepinephrine." You're likely to encounter both in textbooks and exams.

Thriving with the thyroid

The largest of the endocrine glands, the thyroid is like a large butterfly with two lobes connected by a fleshy *isthmus* positioned in the front of the neck, just below the larynx and on either side of the trachea. A transport mechanism called the *iodide pump* moves the iodides from the bloodstream for use in creating its two primary hormones, *thyroxin* and *triiodothyronine,* which regulate the body's metabolic rate. *Extrafollicular cells* (also called *parafollicular* or *C cells*) secrete *calcitonin,* a polypeptide hormone that helps regulate the concentration of calcium and phosphate ions by inhibiting the rate at which they leave the bones. High blood calcium levels stimulate the secretion of more calcitonin.

Thyroxin (T_4) and triiodothyronine (T_3) regulate cellular metabolism throughout the body, but the thyroid needs iodine to manufacture those hormones. Iodine insufficiency causes the thyroid to swell in a condition called a *goiter.*

Pairing up with the parathyroid

The parathyroid consists of four pea-sized glands that lie posterior to the thyroid gland secreting *parathormone,* or *parathyroid hormone* (PTH). This large polypeptide regulates the balance of calcium levels in the blood and bones as well as controls the rate at which calcium is excreted into urine. When blood calcium levels dip, the parathyroid secretes PTH, which increases calcium absorption from the intestine, decreases calcium excretion, increases phosphate excretion, removes calcium from the bones, and stimulates secretion of calcitonin by the thyroid C cells. Blood calcium ion homeostasis is critical to the conduction of nerve impulses, muscle contraction, and blood clotting.

Pinging the pineal gland

The pineal gland, also called the *epiphysis,* is a small, oval gland thought to play a role in regulating the body's biological clock. It lies between the cerebral hemispheres and is attached to the thalamus near the roof of its third ventricle.

Because it both secretes a hormone and receives visual nerve stimuli, the pineal gland is considered part of both the nervous system and the endocrine system. Its hormone *melatonin* is believed to play a role in circadian rhythms, the pattern of repeated behavior associated with the cycles of night and day. The pineal gland is affected by changes in light, producing its highest levels of secretion at night and its lowest levels during daylight hours.

Thumping the thymus

As discussed in Chapter 11, the thymus is thought to secrete a group of peptides called *thymosin* that affect the production of lymphocytes (white blood cells). Thymosin promotes the production and maturation of T lymphocyte cells as part of the body's immune system. The gland is large in children and atrophies with age.

Pressing the pancreas

The pancreas is both an exocrine and an endocrine gland, which means that it secretes some substances through ducts while others go directly into the bloodstream. (We cover its exocrine functions in Chapter 9.) The pancreatic endocrine glands are clusters of cells called the *islets of Langerhans*. Within the islets are a variety of cells, including

- ✔ **A cells (alpha cells)** that secrete the hormone *glucagon*, a polypeptide of 29 amino acids that increases blood sugar

- ✔ **B cells (beta cells)** that secrete *insulin*, a two-linked polypeptide chain of 21 amino acids that decreases blood sugar levels, increases lipid synthesis, and stimulates protein synthesis

- ✔ **D cells (delta cells)** that secrete *somatostatin*, a growth hormone–inhibiting factor that inhibits the secretion of insulin and glucagons

- ✔ **F cells (PP cells)** that secrete a pancreatic polypeptide that regulates the release of pancreatic digestive enzymes

See if all this information has your hormones raging:

15.–19. Mark the statement with a T if it's true or an F if it's false:

15. _____ The adrenal glands are located in the cortex of the kidneys.

16. _____ Adrenaline is functional in the absorption of stored carbohydrates and fat.

17. _____ Aldosterone is functional in regulating the amount of insulin in the body.

18. _____ The sympathetic division of the autonomic nervous system controls the cells of the adrenal medulla.

19. _____ The layers of the adrenal medulla form outer, middle, and inner zones.

20. The endocrine gland that initiates antibody development by producing thymosin is the

a. Pineal body

b. Pituitary gland

c. Thymus

d. Hypothalamus

e. Adrenal gland

21. The hormone that regulates the amount of electrolytes retained in the body is

a. Aldosterone

b. Cortisone

c. Epinephrine

d. Androgens

e. Norepinephrine

22.–26. Mark the statement with a T if it's true or an F if it's false:

22. _____ Iodine is a necessary component of thyroxin (T$_4$) and triiodothyronine (T$_3$).

23. _____ Follicular cells of the thyroid produce hormones that affect the metabolic rate of the body.

24. _____ A transport mechanism called the sodium pump moves the iodides into the follicle cells.

25. _____ Thyroxin (T$_4$) is normally secreted in lower quantity than triiodothyronine (T$_3$).

26. _____ The hormone calcitonin helps regulate the concentration of sodium and potassium.

27. Which statement is *not* true of the pineal gland?

a. It secretes melatonin.

b. Nerve fibers stimulate the pineal cells.

c. As light decreases, secretion increases.

d. It's a small, oval gland.

e. It promotes immunity.

28. Insufficiency of iodine causes the thyroid gland to enlarge, causing

a. Dwarfism

b. Diabetes

c. Giantism

d. Acromegaly

e. Simple or endemic goiter

29.–33. Mark the statement with a T if it's true or an F if it's false:

29. _____ The parathyroid gland contains cells that secrete parathormone or parathyroid hormone (PTH).

30. _____ Melatonin is a polypeptide that regulates the balance of calcium in the blood and bones.

31. _____ The pineal gland responds to light, producing higher levels of secretions at night than during the day.

32. _____ Thymosin promotes the production and maturation of erythrocyte cells.

33. _____ The parathyroid hormone can prompt calcium to move from bone.

34. The endocrine gland that produces 80 percent epinephrine is the

a. Hypothalamus

b. Pituitary

c. Medulla of the adrenal

d. Thyroid

e. Thymus

35. The endocrine gland associated with metabolic rate is the

   a. Parathyroid

   b. Thyroid

   c. Pineal

   d. Posterior lobe of the pituitary

   e. Thymus

Dealing with Stress: Homeostasis

Nothing upsets your delicate cells more than a change in their internal environment. A stimulus such as fear or pain provokes a response that upsets your body's carefully maintained equilibrium. Such a change initiates a nerve impulse to the hypothalamus that activates the sympathetic division of the autonomic nervous system and increases secretions from the adrenal glands. This change — called a *stressor* — produces a condition many know oh so well: *stress*. The body's immediate response is to push for *homeostasis* — keeping everything the same inside.

The body's effort to maintain homeostasis invokes a series of reactions called the *general stress syndrome* that's controlled by the hypothalamus. When the hypothalamus receives stress information, it responds by preparing the body for fight or flight; in other words some kind of decisive, immediate, physical action. This reaction increases blood levels of glucose, glycerol, and fatty acids; increases the heart rate and breathing rate; redirects blood from skin and internal organs to the skeletal muscles; and increases the secretion of adrenaline from the adrenal medulla. The hypothalamus releases *corticotropin-releasing hormone* (CRH) that stimulates the anterior lobe of the pituitary to secrete *adrenocorticotropic hormone* (ACTH), which tells the adrenal cortex to secrete more cortisone. That cortisone supplies the body with amino acids and an extra energy source needed to repair any injured tissues that may result from the impending crisis.

As part of the general stress syndrome, the pancreas produces glucagon, and the anterior pituitary secretes growth hormones, both of which prepare energy sources and stimulate the absorption of amino acids to repair damaged tissue. The posterior pituitary secretes antidiuretic hormone, making the body hang on to sodium ions and spare water. The subsequent decrease in urine output is important to increase blood volume, especially if there's bleeding or excessive sweating.

Wow. With the body gearing up like that every time, it's no wonder that people subjected to repeated stress are often sickly.

We try not to stress you out with these practice questions:

36.–40. Mark the statement with a T if it's true or an F if it's false:

   36. _____ The hypothalamus controls reactions to combat general stress syndrome.

   37. _____ The pancreas is an endocrine gland only.

   38. _____ During stress, the pancreas produces thyroxin (T_4).

   39. _____ Alpha cells in the pancreas secrete the hormone insulin.

   40. _____ Changes in the body's environment called stressors produce a condition called stress.

41. When changes occur in the body's internal environment, a reaction is initiated by

    a. Neurohormones

    b. Glucocorticoids

    c. The hypothalamus

    d. The adrenal cortex

    e. The pituitary gland

42. Stress activates a set of body responses called

    a. The survival response

    b. The general stress syndrome

    c. The repair response

    d. The resistance response

    e. The stress reflex

43. The body's initial reaction to a stressor is

    a. Fight or flight response

    b. Repair response

    c. To promote rapid wound healing

    d. Stress reflex

    e. To promote normal metabolism

44. Which of the following is a response to stress?

    a. Decrease the heart rate

    b. Increase the urine output

    c. Redirect blood from the skeletal muscles

    d. Increase the respiratory rate

    e. Decrease the glucose in the blood

45. The pancreas, testes, and ovaries all have this in common:

    a. All are influenced by hormones from the parathyroid.

    b. All are considered to be both exocrine and endocrine.

    c. None were formed from embryonic tissues.

    d. They influence secondary sex characteristics.

    e. They have no blood supply.

46.–55. Use the terms that follow to identify the structures of the endocrine system shown in Figure 16-2:

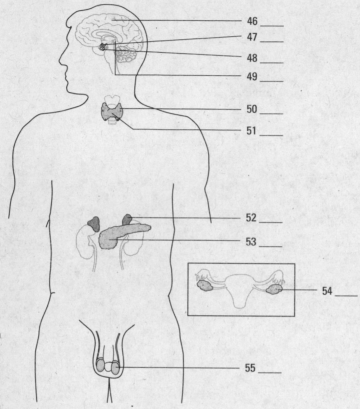

46 ____
47 ____
48 ____
49 ____
50 ____
51 ____
52 ____
53 ____
54 ____
55 ____

Figure 16-2:
The
endocrine
system.

LifeART Image Copyright © 2007. Wolters Kluwer Health — Lippincott Williams & Wilkins

a. Thyroid gland

b. Pineal gland

c. Pituitary gland

d. Adrenal gland

e. Ovaries

f. Parathyroid gland

g. Testes

h. Hypothalamus

i. Pancreas

j. Brain

Answers to Questions on the Endocrine System

The following are answers to the practice questions presented in this chapter.

1. The endocrine system brings about changes in the metabolic activities of the body tissue. **True.** Metabolism is one of the areas influenced by hormones.

2. The amount of hormone released is determined by the body's need for that hormone at the time. **True.** In many ways, it's a self-regulating system: Just enough hormones are distributed to balance everything else out.

3. The glands of the endocrine system are composed of cartilage cells. **False.** That connection makes no sense whatsoever.

4. Endocrine glands aren't functional in reproductive processes. **False.** The endocrine system is a key component in reproduction.

5. Some hormones can be derivatives of amino acids, whereas others are synthesized from cholesterol. **True.** Amino acids for the nonsteroids and cholesterol for the steroid-based hormones.

6. Glands that secrete their product into the interstitial fluid, which flows into the blood, are **b. endocrine glands.**

7. Cells that respond to a hormone are **c. target cells.** Hormones actually go looking for these specific targets.

8. The pituitary gland consists of two parts: an endocrine gland and modified nerve tissue. **True.** Remember that the anterior lobe is mostly epithelial cells, whereas the posterior lobe contains primarily nerve cells.

9. The pituitary gland is found in the sella turcica of the temporal bone. **False.** The pituitary gland is in the sphenoid bone, not the temporal bone.

10. The adenohypophysis is called the master gland because of its influence on all the body's tissues. **False.** It earned the title "master gland" because of its influence over the other endocrine glands.

11. ADH causes constriction of smooth muscle tissue in the blood vessels, which elevates the blood pressure. **True.**

12. The neurohypophysis stores and releases secretions produced by the hypothalamus. **True.** Seems a strange thing for a structure made of nerve cells to do, but it does its job well.

13. The gland that does the most to regulate and maintain the function of other glands is the **b. pituitary.** That's why it's the master gland.

14. Which of the following is not a pituitary hormone? **a. Progesterone.** That's made by the corpus luteum on the ovary following ovulation.

15. The adrenal glands are located in the cortex of the kidneys. **False.** They're atop the kidneys.

16. Adrenaline is functional in the absorption of stored carbohydrates and fat. **False.** Adrenaline does lots of things, but not that.

17   Aldosterone is functional in regulating the amount of insulin in the body. **False.** This hormone regulates mineral salts.

18   The sympathetic division of the autonomic nervous system controls the cells of the adrenal medulla. **True.**

19   The layers of the adrenal medulla form outer, middle, and inner zones. **False.** The layers of the adrenal cortex form those three zones.

20   The endocrine gland that initiates antibody development by producing thymosin is the **c. thymus.**

TIP   The name of the hormone, thymosin, should be your first clue that it's produced by the thymus. See the resemblance?

21   The hormone that regulates the amount of electrolytes retained in the body is **a. aldosterone.**

22   Iodine is a necessary component of thyroxin (T_4) and triiodothyronine (T_3). **True.** The body can't make those hormones without iodine.

23   Follicular cells of the thyroid produce hormones that affect the metabolic rate of the body. **True.**

24   A transport mechanism called the sodium pump moves the iodides into the follicle cells. **False.** Don't let the thyroid's iodide pump make you think it also has a sodium pump. It doesn't.

25   Thyroxin (T_4) is normally secreted in lower quantity than triiodothyronine (T_3). **False.** In fact, it's just the opposite — more T_4 is secreted than T_3.

26   The hormone calcitonin helps regulate the concentration of sodium and potassium. **False.** Actually, calcitonin lowers plasma calcium and phosphate levels.

27   Which statement is *not* true of the pineal gland? **e. It promotes immunity.** It's more of a biological-clock kind of gland.

28   Insufficiency of iodine causes the thyroid gland to enlarge, causing **e. simple or endemic goiter.** Sometimes this swelling becomes visible at the base of the neck.

29   The parathyroid gland contains cells that secrete parathormone or parathyroid hormone (PTH). **True.**

30   Melatonin is a polypeptide that regulates the balance of calcium in the blood and bones. **False.** Melatonin is thought to regulate circadian rhythms.

31   The pineal gland responds to light, producing higher levels of secretions at night than during the day. **True.** That's why it's also considered part of the nervous system.

32   Thymosin promotes the production and maturation of erythrocyte cells. **False.** Thymosin works on lymphocytes.

33   The parathyroid hormone can prompt calcium to move from bone. **True.** After all, the bones are mineral reservoirs.

34   The endocrine gland that produces 80 percent epinephrine is the **c. medulla of the adrenal.** That's about all the adrenal medulla does.

35 The endocrine gland associated with metabolic rate is the **b. thyroid.** Controlling the body's metabolic rate is its primary role.

36 The hypothalamus controls reactions to combat general stress syndrome. **True.**

37 The pancreas is an endocrine gland only. **False.** It's also an exocrine gland.

38 During stress, the pancreas produces thyroxin (T_4). **False.** That's the thyroid's job, and T_4 has nothing to do with stress, anyway.

39 Alpha cells in the pancreas secrete the hormone insulin. **False.** A cells secrete glucagon.

40 Changes in the body's environment called stressors produce a condition called stress. **True.**

41 When changes occur in the body's internal environment, a reaction is initiated by **c. the hypothalamus.** It stimulates the sympathetic division of the autonomic nervous system and adrenal medulla.

42 Stress activates a set of body responses called **b. the general stress syndrome.**

43 The body's initial reaction to a stressor is **a. fight or flight response.** You're either going to put up your dukes or run like the wind.

44 Which of the following is a response to stress? **d. Increase the respiratory rate.** That brings in more oxygen in case you need to move fast.

45 The pancreas, testes, and ovaries all have this in common: **b. All are considered to be both exocrine and endocrine.** Secretions go directly into the bloodstream *and* through ducts.

46–55 Following is how Figure 16-2, the endocrine system, should be labeled.

46 **j. Brain**; 47. **h. Hypothalamus**; 48. **c. Pituitary gland**; 49. **b. Pineal gland**; 50. **f. Parathyroid gland**; 51. **a. Thyroid gland**; 52. **d. Adrenal gland**; 53. **i. Pancreas**; 54. **e. Ovaries**; 55. **g. Testes**

Part VI
The Part of Tens

The 5th Wave   By Rich Tennant

IN THE PHYSIOLOGY DEPT., PROFESSOR RICKMAN WAS KNOWN AS BEING ONE. LOBE. SHORT OF A LIMBIC SYSTEM.

BOINK!

In this part . . .

This classic *For Dummies* feature contains two chapters identifying ten useful tidbits for students of anatomy and physiology. We identify ten Web sites that can help you advance your knowledge and ten top study tips to help you get A's (or at least get closer to them!) in anatomy and physiology courses.

Chapter 17

Ten Study Tips

What's the best way to tackle anatomy and physiology and come out successful on the other side? Of course, a good memory helps, but it's not as critical to anatomy and physiology success as it's cracked up to be. With a little advance planning and tricks of the study trade, even students who complain that they can't remember their own names on exam day can summon the right terminology and information from their scrambled synaptic pathways. In this chapter, we cover ten key things you can start doing today to ensure success not only in anatomy and physiology but in any number of other classes.

Write It Down in Your Own Words

This is a simple idea that far too few students practice regularly. Don't stop at underlining and highlighting important material in your textbooks and study guides: Write it down. Or type it up. Whatever you do, don't just regurgitate it exactly as presented in the material you're studying. Find your own words. Create your own analogies. Tell your own tale of what happens to the bolus as it ventures into the digestive tract. Detail the course followed by a molecule of oxygen as it enters through the nose. Draw pictures of the differences between meiosis and mitosis. Completely relax into the process with the knowledge that no one else ever has to see what you write, type, or sketch. All they'll ever see is your successful completion of the course!

Better Knowledge through Mnemonics

Studying anatomy and physiology involves remembering lists of terms, functions, and processes. Sprinkled throughout this book are suggestions to take just the first letter or two of each word from a list to create an acronym. Occasionally, we help you go one step beyond the acronym to a clever little thing called a *mnemonic device*. Simply put, the mnemonic is the thing you commit to memory as a means for remembering the more technical thing for which it stands. For example, a question in Chapter 7 asks you to list in order the epidermal layers from the dermis outward; we suggest that you commit the following phrase to memory: Be Super Greedy, Less Caring. Just like that, a complicated list like basale, spinosum, granulosum, lucidum, corneum gets a little closer to a permanent home in your brain.

Not feeling terribly clever at the moment you need a useful mnemonic? Surf on over to www.medicalmnemonics.com, which touts itself as the world's database of these useful tools. Here's a sampling of the site's offerings:

- To remember the three types of tonsils: "PPL (people) have tonsils: **P**haryngeal, **P**alatine, and **L**ingual."

- To remember the spleen's location and dimensions: "Count 1, 3, 5, 7, 9, 11: The spleen is 1 inch by 3 inches by 5 inches, weighs 7 ounces, and underlies ribs 9 through 11."

- To remember the cranial bones; "Think PEST OF 6: **P**arietal, **E**thmoid, **S**phenoid, **T**emporal, **O**ccipital, **F**rontal."

Stylish Learning

Every person has his or her own sense of style, and woe betide anyone who tries to shoehorn the masses into a single style. The same, of course, is true of students. To get the most out of your study time, you need to figure out what your learning style is and alter your study habits to accommodate it. No idea what we're talking about? Answer the questionnaire posted at www.vark-learn.com/english/page.asp?p=questionnaire, and the VARK guide to learning styles will tell you more about yourself than your last psychotherapist.

VARK, as you may have suspected, is an acronym that stands for Visual (learning by seeing), Aural (learning by hearing), Reading/Writing (learning by reading and writing), and Kinesthetic (learning by touching, holding, or feeling). If you're a visual learner, you may get more out of anatomy and physiology by seeing the real thing in the flesh. If you're an aural learner, you may learn best in the classroom as the teacher lectures. If you're a reading and writing kind of learner, you'll get the most out of our first tip to write stuff down. And if you're a kinesthetic learner, there's nothing like touching or holding to commit something to memory.

Grecian Formula

If you keep thinking "It's all Greek to me," congratulations on your insight! The truth of the matter is that most of it actually *is* Greek. So dust off your foreign language learning skills and begin with the basic vocabulary of medical terminology. (Get started with the Greek and Latin roots, prefixes, and suffixes that appear on this book's Cheat Sheet — the tear-out page in the front of the book.) You'll soon discover that for every "little" word you learn, there's a whole mountain of additional terms and phrases just waiting to be discovered.

Connecting with Concepts

It happens time and again in anatomy and physiology that one concept or connection mirrors another yet to be learned. But because you're focusing so hard on this week's lesson, you lose sight of the value in the previous month's lessons. For example, a concept like metabolism comes up in a variety of ways throughout your study of anatomy and physiology. When you encounter a repeat concept like that, create a special page or two for it at the back of your notebook, or link the concept to a separate computer file. Then, every time the term comes up in class or in your textbook, add to the running list of notes on that concept. You'll have references to metabolism at each point it comes up *and* you'll be able to analyze its influences across different body systems.

Grouping Studies

If you're really lucky, someone in your class (or maybe it's even you) has already suggested forming that time-honored tradition — a *study group*. The power of group members to fill gaps in your knowledge is priceless. But don't restrict it to late-night cramming just before each test. Meet with your group at least once a week to go over lecture notes and textbook readings. If it's true that people only retain about 10 percent of what they hear or read, then it makes sense that your fellow group members will recall things that slipped immediately from your mind.

Outlining What's to Come

As you read through a chapter of your textbook to prepare for the next lecture, prepare an outline of what you're reading, leaving plenty of space between subheadings. Then, during the lecture, take your notes within the outline you've already created. Piecing together an incomplete puzzle shows you where the key gaps in your knowledge may be.

Practice, Practice, Practice

Flash cards, mnemonic drills, practice tests — be creative and practice, practice, practice! The more you know about the format of any upcoming exam, the better. Sometimes instructors share tidbits about what they plan to emphasize, but sometimes they don't. In the end, if you've done the work and put in the time to study and practice with information outside of class, the exact structure and content of an exam shouldn't make much difference.

Sleuthing Out Clues

Okay, it's test time! Take advantage of the test itself. You may find that the answer to an exam question that stumps you is revealed — at least partially — in the phrasing of a subsequent question. Stay alert to these blessed little gifts even when you think that you already understand the whole anatomical process. You wouldn't be the first student to change an answer after working your way through an exam.

Learning from Mistakes

The test is done and the grades are in. So there was a really tough question or two on the test and you blew it big-time? It's hardly a missed opportunity — this is where rolling with the punches really pays off. Go back over the entire test and pay extra attention to what you got wrong. Start your next practice sessions with those questions, and stay alert for upcoming material that may trip you up in a similar way.

Chapter 18

Ten (Plus One) Terrific Online Resources

In This Chapter
▶ Scanning online reference books
▶ Exploring databases and virtual tours

No matter how much you study or how many Latin and Greek roots you memorize, it's inevitable that some aspects of anatomy and physiology will leave you dazed and confused. But if you study within reach of an Internet connection, you don't have to stay that way for long. Simply surf over to one of the 11 sites covered in this chapter and start entering search terms. As with anything Internet-related, however, you have to be cautious about the accuracy of what you find. Just keep our mantra in mind: "When in doubt, trust the textbook."

Answers and More Answers

www.answers.com

Admittedly, this free Web site doesn't focus on anatomy and physiology, but you won't really care about that technicality after you've dived into it a few times. The site offers about 4 million "answers" based more on keyword searches than on any actual questions. Material is drawn from scores of brand-name content publishers as well as Answers.com's own editorial team. Can't quite figure out where an "anteroinferior" something is supposed to be? Forgot where Peyer's patches are hiding? Not entirely sure what a gallbladder does? Answers.com's friendly little "Tell me about . . ." box at the top of the home page takes you straight to a Web page that aggregates what various sources say on your chosen topic.

Into the Lion's Den

www.lionden.com/ap.htm

We told you to trust the textbook, didn't we? That means that you can probably trust the textbook's author, too. This delightful site is maintained by a Missouri community college professor who also happens to be coauthor of an anatomy and physiology textbook. Dr. Kevin T. Patton has been teaching the subject for more than two decades and has developed a refreshingly gentle sense of humor along the way. His Web site is packed with study tips, downloadable PowerPoint slides, and guided tours of various anatomical systems.

The Venerable Gray's

www.bartleby.com/107

When we refer to "Gray's Anatomy," we're not talking about the popular TV series of a similar name. We're talking about the venerable reference book that dates back to 1858 and is now online for quick and easy access. Included in the "virtual" Gray's *Anatomy of the Human Body* are more than 1,200 color illustrations and a subject index with 13,000 entries. If your textbook is missing critical illustrations, don't worry; a quick keyword search on this Web site can reveal a number of relevant graphics for you to study.

Alluring Anatomists

www.anatomy.org

The American Association of Anatomists believes that anatomy is "truly the backbone of biomedical science." With that in mind, the organization has put together an Ask the Expert feature that lets anyone — students and educators alike — query the group's working professionals. To access this feature, click the Education and Teaching Tools link on the organization's home page; then select Ask the Expert. (To view the answer to a particular posted question, click on the question.)

From the Education and Teaching Tools page, you also can find a list of popular links to various subspecialties, including cell biology, genetics, imaging, molecular development, endocrinology, forensics, and physical anthropology.

Getting Body Smart

www.getbodysmart.com

This Web site is the brainchild and passion of an anatomy and physiology instructor, Scott Sheffield, who says in his site's mission statement that he's attempting to distill two decades of teaching into a single, fully animated and interactive e-book about the human body. In addition to "flash" windows that drill down into various systems, GetBodySmart offers free tutorials and quizzes to explain complex physiological interactions. Sheffield readily acknowledges that his work will last "many years," so perhaps the best is yet to come.

Pop Quiz Central

msjensen.education.umn.edu/webanatomy

Murray Jensen, an associate professor at the University of Minnesota, conducts research on the use of technology in science education. His Web site is both an outgrowth of his research and a source of ideas for it. As of this writing, the site consists mostly of dozens of quizzes of varying lengths and difficulty. Treat this site like your own personal flash card system and you'll be head and shoulders above your fellow students.

MEDTropolis – Virtual Body

www.medtropolis.com/VBody.asp

When you visit this site, click on English or Spanish, and then sit back and enjoy the show. This site only covers the brain, skeleton, heart, and digestive tract, but its clear, concise three-dimensional representations of these organs and systems make it worth a look.

Drilling, Drilling, Drilling Some More

matcmadison.edu/faculty/cshuster/wiley.html

This Web site links to an extensive set of anatomy drill and practice exercises maintained by John Wiley and Sons (who also happens to be the publisher of this book). To work through the multiple-choice practice questions, you need to download the Shockwave plug-in (if you don't already have it), but that's a small price to pay for such a useful site. Full-color images with blank labels give you the opportunity to figure out which part is what and why; then you can clear your labels and begin again as often as you like.

Human Biodyssey: Exploring Anatomy and Physiology

www.gwc.maricopa.edu/home_pages/crimando/jcHumanBiodyssey.htm

This page is the impressive work of Dr. James Crimando at Gateway Community College in Phoenix, Arizona. Dr. Crimando lists every scrap of information a student needs to succeed in his classes (or any anatomy and physiology class), including extensive practice questions, lecture outlines, and quick summaries of class sessions. Regardless of whether you're among Dr. Crimando's students, his site is an incredibly useful receptacle for information about how the body is organized.

List of Lists

www.mhhe.com/biosci/ap/saladin/www.mhtml

Textbook author Kenneth Saladin uses a small portion of this site to provide descriptions of the last three versions of his anatomy and physiology textbooks. But that doesn't begin to compare with what he has done pulling together resources from all over the Web in the Student Resources page we guide you to here. So much to see, so little time!

Virtual Anatomy

library.thinkquest.org/16421/noframes/index.htm

Believe it or not, this site was created by a trio of high school kids in San Antonio, Texas. The site's capabilities are somewhat limited, but it contains some good interactive anatomical practice areas and a couple of educational videos, too.

Index

BUSINESS, CAREERS & PERSONAL FINANCE

0-7645-9847-3

0-7645-2431-3

Also available:
- Business Plans Kit For Dummies
  0-7645-9794-9
- Economics For Dummies
  0-7645-5726-2
- Grant Writing For Dummies
  0-7645-8416-2
- Home Buying For Dummies
  0-7645-5331-3
- Managing For Dummies
  0-7645-1771-6
- Marketing For Dummies
  0-7645-5600-2

- Personal Finance For Dummies
  0-7645-2590-5*
- Resumes For Dummies
  0-7645-5471-9
- Selling For Dummies
  0-7645-5363-1
- Six Sigma For Dummies
  0-7645-6798-5
- Small Business Kit For Dummies
  0-7645-5984-2
- Starting an eBay Business For Dummies
  0-7645-6924-4
- Your Dream Career For Dummies
  0-7645-9795-7

HOME & BUSINESS COMPUTER BASICS

0-470-05432-8

0-471-75421-8

Also available:
- Cleaning Windows Vista For Dummies
  0-471-78293-9
- Excel 2007 For Dummies
  0-470-03737-7
- Mac OS X Tiger For Dummies
  0-7645-7675-5
- MacBook For Dummies
  0-470-04859-X
- Macs For Dummies
  0-470-04849-2
- Office 2007 For Dummies
  0-470-00923-3

- Outlook 2007 For Dummies
  0-470-03830-6
- PCs For Dummies
  0-7645-8958-X
- Salesforce.com For Dummies
  0-470-04893-X
- Upgrading & Fixing Laptops For Dummies
  0-7645-8959-8
- Word 2007 For Dummies
  0-470-03658-3
- Quicken 2007 For Dummies
  0-470-04600-7

FOOD, HOME, GARDEN, HOBBIES, MUSIC & PETS

0-7645-8404-9

0-7645-9904-6

Also available:
- Candy Making For Dummies
  0-7645-9734-5
- Card Games For Dummies
  0-7645-9910-0
- Crocheting For Dummies
  0-7645-4151-X
- Dog Training For Dummies
  0-7645-8418-9
- Healthy Carb Cookbook For Dummies
  0-7645-8476-6
- Home Maintenance For Dummies
  0-7645-5215-5

- Horses For Dummies
  0-7645-9797-3
- Jewelry Making & Beading For Dummies
  0-7645-2571-9
- Orchids For Dummies
  0-7645-6759-4
- Puppies For Dummies
  0-7645-5255-4
- Rock Guitar For Dummies
  0-7645-5356-9
- Sewing For Dummies
  0-7645-6847-7
- Singing For Dummies
  0-7645-2475-5

INTERNET & DIGITAL MEDIA

0-470-04529-9

0-470-04894-8

Also available:
- Blogging For Dummies
  0-471-77084-1
- Digital Photography For Dummies
  0-7645-9802-3
- Digital Photography All-in-One Desk Reference For Dummies
  0-470-03743-1
- Digital SLR Cameras and Photography For Dummies
  0-7645-9803-1
- eBay Business All-in-One Desk Reference For Dummies
  0-7645-8438-3
- HDTV For Dummies
  0-470-09673-X

- Home Entertainment PCs For Dummies
  0-470-05523-5
- MySpace For Dummies
  0-470-09529-6
- Search Engine Optimization For Dummies
  0-471-97998-8
- Skype For Dummies
  0-470-04891-3
- The Internet For Dummies
  0-7645-8996-2
- Wiring Your Digital Home For Dummies
  0-471-91830-X

* Separate Canadian edition also available
† Separate U.K. edition also available

Available wherever books are sold. For more information or to order direct: U.S. customers visit www.dummies.com or call 1-877-762-2974.
U.K. customers visit www.wileyeurope.com or call 0800 243407. Canadian customers visit www.wiley.ca or call 1-800-567-4797.

SPORTS, FITNESS, PARENTING, RELIGION & SPIRITUALITY

0-471-76871-5

0-7645-7841-3

Also available:
- Catholicism For Dummies
  0-7645-5391-7
- Exercise Balls For Dummies
  0-7645-5623-1
- Fitness For Dummies
  0-7645-7851-0
- Football For Dummies
  0-7645-3936-1
- Judaism For Dummies
  0-7645-5299-6
- Potty Training For Dummies
  0-7645-5417-4
- Buddhism For Dummies
  0-7645-5359-3

- Pregnancy For Dummies
  0-7645-4483-7 †
- Ten Minute Tone-Ups For Dummies
  0-7645-7207-5
- NASCAR For Dummies
  0-7645-7681-X
- Religion For Dummies
  0-7645-5264-3
- Soccer For Dummies
  0-7645-5229-5
- Women in the Bible For Dummies
  0-7645-8475-8

TRAVEL

0-7645-7749-2

0-7645-6945-7

Also available:
- Alaska For Dummies
  0-7645-7746-8
- Cruise Vacations For Dummies
  0-7645-6941-4
- England For Dummies
  0-7645-4276-1
- Europe For Dummies
  0-7645-7529-5
- Germany For Dummies
  0-7645-7823-5
- Hawaii For Dummies
  0-7645-7402-7

- Italy For Dummies
  0-7645-7386-1
- Las Vegas For Dummies
  0-7645-7382-9
- London For Dummies
  0-7645-4277-X
- Paris For Dummies
  0-7645-7630-5
- RV Vacations For Dummies
  0-7645-4442-X
- Walt Disney World & Orlando
  For Dummies
  0-7645-9660-8

GRAPHICS, DESIGN & WEB DEVELOPMENT

0-7645-8815-X

0-7645-9571-7

Also available:
- 3D Game Animation For Dummies
  0-7645-8789-7
- AutoCAD 2006 For Dummies
  0-7645-8925-3
- Building a Web Site For Dummies
  0-7645-7144-3
- Creating Web Pages For Dummies
  0-470-08030-2
- Creating Web Pages All-in-One Desk
  Reference For Dummies
  0-7645-4345-8
- Dreamweaver 8 For Dummies
  0-7645-9649-7

- InDesign CS2 For Dummies
  0-7645-9572-5
- Macromedia Flash 8 For Dummies
  0-7645-9691-8
- Photoshop CS2 and Digital
  Photography For Dummies
  0-7645-9580-6
- Photoshop Elements 4 For Dummies
  0-471-77483-9
- Syndicating Web Sites with RSS Feeds
  For Dummies
  0-7645-8848-6
- Yahoo! SiteBuilder For Dummies
  0-7645-9800-7

NETWORKING, SECURITY, PROGRAMMING & DATABASES

0-7645-7728-X

0-471-74940-0

Also available:
- Access 2007 For Dummies
  0-470-04612-0
- ASP.NET 2 For Dummies
  0-7645-7907-X
- C# 2005 For Dummies
  0-7645-9704-3
- Hacking For Dummies
  0-470-05235-X
- Hacking Wireless Networks
  For Dummies
  0-7645-9730-2
- Java For Dummies
  0-470-08716-1

- Microsoft SQL Server 2005 For Dummies
  0-7645-7755-7
- Networking All-in-One Desk Reference
  For Dummies
  0-7645-9939-9
- Preventing Identity Theft For Dummies
  0-7645-7336-5
- Telecom For Dummies
  0-471-77085-X
- Visual Studio 2005 All-in-One Desk
  Reference For Dummies
  0-7645-9775-2
- XML For Dummies
  0-7645-8845-1